**Pulse Oxime**

Dilpreet Kaur

# Pulse Oximeter using ADuC842 Microcontroller

## A monitoring device for measuring blood oxygen saturation and pulse rate

LAP LAMBERT Academic Publishing

**Impressum/Imprint (nur für Deutschland/only for Germany)**
Bibliografische Information der Deutschen Nationalbibliothek: Die Deutsche Nationalbibliothek verzeichnet diese Publikation in der Deutschen Nationalbibliografie; detaillierte bibliografische Daten sind im Internet über http://dnb.d-nb.de abrufbar.
Alle in diesem Buch genannten Marken und Produktnamen unterliegen warenzeichen-, marken- oder patentrechtlichem Schutz bzw. sind Warenzeichen oder eingetragene Warenzeichen der jeweiligen Inhaber. Die Wiedergabe von Marken, Produktnamen, Gebrauchsnamen, Handelsnamen, Warenbezeichnungen u.s.w. in diesem Werk berechtigt auch ohne besondere Kennzeichnung nicht zu der Annahme, dass solche Namen im Sinne der Warenzeichen- und Markenschutzgesetzgebung als frei zu betrachten wären und daher von jedermann benutzt werden dürften.

Coverbild: www.ingimage.com

Verlag: LAP LAMBERT Academic Publishing GmbH & Co. KG
Heinrich-Böcking-Str. 6-8, 66121 Saarbrücken, Deutschland
Telefon +49 681 3720-310, Telefax +49 681 3720-3109
Email: info@lap-publishing.com

Approved by: Gurgaon, ITM University, Diss., 2011

Herstellung in Deutschland (siehe letzte Seite)
ISBN: 978-3-8484-9318-0

**Imprint (only for USA, GB)**
Bibliographic information published by the Deutsche Nationalbibliothek: The Deutsche Nationalbibliothek lists this publication in the Deutsche Nationalbibliografie; detailed bibliographic data are available in the Internet at http://dnb.d-nb.de.
Any brand names and product names mentioned in this book are subject to trademark, brand or patent protection and are trademarks or registered trademarks of their respective holders. The use of brand names, product names, common names, trade names, product descriptions etc. even without a particular marking in this works is in no way to be construed to mean that such names may be regarded as unrestricted in respect of trademark and brand protection legislation and could thus be used by anyone.

Cover image: www.ingimage.com

Publisher: LAP LAMBERT Academic Publishing GmbH & Co. KG
Heinrich-Böcking-Str. 6-8, 66121 Saarbrücken, Germany
Phone +49 681 3720-310, Fax +49 681 3720-3109
Email: info@lap-publishing.com

Printed in the U.S.A.
Printed in the U.K. by (see last page)
ISBN: 978-3-8484-9318-0

# ACKNOWLEDGEMENTS

The completion of this book is a fruit of a combined work. Therefore, the sense of satisfaction that accompanies it also has to be shared with all those generous persons, who were instrumental in bringing this work in a reality.

First and foremost, I would like to express my gratitude to my supervisors, Dr. Amod Kumar, Mrs Shashi Sharma and Mr. Sukhwinder Kumar for their continuous encouragement and support. Special thanks to all the teachers of EECE department, ITM University, Gurgaon for their assistance and making things easier for me.

The long journey was possible due to the friendly research environment. I would also like to thank my friends who have helped me at different stages of research and for their insightful comments.

I thank my parents for their love, care, protection and having confidence in my abilities. They were with me at each and every step and encouraged me not to give up at tough times and sustained me all through. My brother, Harsimran who always inspired me to achieve success and always supported me in everything I did.

It is with same sense of gratitude that I recall the support of my uncle, Mr. Jasbir Singh and my aunt Ms. Hardeep Kaur, who always have high expectations from me and encouraged me to achieve the best.

At last but not the least, I would like to thank my grandparents; only because of their unconditional love and support, I was able to complete this book.

Dilpreet Kaur
Department of Electrical & Electronics Communication
Amity University
Haryana

To my loving husband
Amandeep Batra

# CONTENTS

**CHAPTER 1: INTRODUCTION**

1.1     History of Pulse Oximeter ............................................................................................1

1.2     General Concepts ........................................................................................................3

1.3     Advantages and Disadvantages of Pulse Oximeter .....................................................5

1.3.1 Advantages of Pulse Oximeter.............................................................................5

1.3.2 Disadvantages of Pulse Oximeter ........................................................................6

1.4     Applications of Pulse Oximeter ...................................................................................7

1.5     Comparison of pulse oximeter with Arterial Blood Gas (ABG) sampling.................8

1.6     Problem Statement .......................................................................................................8

**CHAPTER 2: DETAILS OF TECHNOLOGY USED**

2.1     Principle of pulse oximetry technology.......................................................................9

2.2     Working of pulse oximeter.........................................................................................13

2.3     Design Parameters .....................................................................................................14

2.3.1 Heart Rate ..........................................................................................................14

2.3.2 Blood oxygen saturation ....................................................................................17

2.4     Different types of pulse oximeters ............................................................................18

2.5     Difference between blood and hemoglobin ..............................................................20

**CHAPTER 3: WORK DONE**

3.1     Block Diagram ..........................................................................................................24

3.1.1 PPG(Photoplethysmograph) Probe .....................................................................24

3.1.2 Amplifier Stage ..................................................................................................29

3.1.3 ADC Stage ..........................................................................................................31

3.1.4 Microcontroller ..................................................................................................44

3.1.5 OLED Display......................................................................................................65

**CHAPTER 4: HARDWARE IMPLEMENTATION**

4.1     Circuit diagram.............................................................................................75

4.2     Description of the circuit..............................................................................76

4.2.1 About PICASO-GFX2 microcontroller ...................................................83

4.3     Testing and troubleshooting of the circuit..................................................91

4.4     Obtaining the waveform for the heart beat.................................................91


**CHAPTER 5: SOFTWARE IMPLEMENTATION**

5.1     Programming environment............................................................................93

5.1.1 About the software ..................................................................................93

5.1.2 Compiler and assembler configuration ..................................................93

5.1.3 Managing projects...................................................................................95

5.1.4 Managing files and folders in a project..................................................99

5.1.5 Compiling and Building.........................................................................100

5.2     Software Design ..........................................................................................104

5.2.1 Flowchart for the calculation of blood oxygen saturation and the heart rate........104

5.3     Screen shots.................................................................................................108


**CHAPTER 6: RESULTS**

6.1     Readings .....................................................................................................111

6.2     Results ........................................................................................................111

6.3     Conclusion...................................................................................................112

6.4     Future Scope................................................................................................112

**REFERENCES**...............................................................................................................113

**BIBLIOGRAPHY** .......................................................................................................114

# LIST OF FIGURES

Figure 1-1: Benchtop Pulse Oximeter.................................................................................... 5

Figure 2-1: Absorption relationship of oxygen levels in blood for the red & IR wavelengths ...... 9

Figure 2-2: Transmission and reflectance type of sensor ........................................................ 11

Figure 2-3: The principle difference between transmission and reflectance pulse oximetry ....... 12

Figure 2-4: Light absorption by tissue type ............................................................................ 13

Figure 2-5: Light absorption by tissues.................................................................................. 13

Figure 2-6 Front of right upper extremity............................................................................... 15

Figure 2-7: Arteries of the neck ............................................................................................ 17

Figure 2-8: Finger tip pulse oximeter .................................................................................... 18

Figure 2-9 Handheld pulse oximeter...................................................................................... 19

Figure 2-10: Fetal pulse oximeter ......................................................................................... 19

Figure 2-11 Red blood cell structure...................................................................................... 21

Figure 2-12: Hemoglobin protein and iron-containing heme group ......................................... 22

Figure 2-13: Oxygen-Hemoglobin dissociation curve............................................................. 23

Figure 3-1: Block diagram of pulse oximeter .......................................................................... 24

Figure 3-2: Representative PPG taken from an ear pulse oximeter........................................... 25

Figure 3-3: Layers of human skin........................................................................................... 26

Figure 3-4: Nellcor SpO$_2$ sensor .......................................................................................... 27

Figure 3-5: DB9 connector .................................................................................................... 28

Figure 3-6: Finger-clip with the connector ............................................................................. 29

Figure 3-7: IC UA741 .......................................................................................................... 31

Figure 3-8: ADC Transfer Function........................................................................................ 32

Figure 3-9: ADC result word format....................................................................................... 33

Figure 3-10: Internal ADC structure....................................................................................... 34

Figure 3-11: Buffering analog inputs...................................................................................... 35

Figure 3-12: Decoupling $V_{REF}$ and $C_{REF}$............................................................................. 37

Figure 3-13: Using an external voltage reference .................................................................... 38

Figure 3-14: Functional block diagram of ADuC842 microcontroller ....................................... 45

Figure 3-15: Pin diagram of ADuC842 microcontroller............................................................ 46

Figure 3-16: ADuC842 microcontroller card........................................................................... 50

Figure 3-17: Lower 128 bytes of internal data memory ........................................................... 52

Figure 3-18: Extended stack pointer operation ....................................................................... 53

Figure 3-19: Internal and external XRAM ............................................................................ 54

Figure 3-20: Programming model ...................................................................................... 55

Figure 3-21: Block diagram of TIC ................................................................................... 58

Figure 3-22: Micro-OLED display .................................................................................... 66

Figure 3-23: Circuit diagram of µOLED-3202X-P1 powered by PICASO-GFX ..................... 68

Figure 3-24: Pin diagram of µOLED-3202X-P1 ................................................................. 69

Figure 3-25: Micro-USB interface with the USD card ......................................................... 70

Figure 3-26: USB to serial interface ................................................................................. 74

Figure 4-1: Circuit diagram of pulse oximeter .................................................................... 75

Figure 4-2: ORCAD capture of the circuit diagram ............................................................. 76

Figure 4-3: Pin diagram of IC 7404 .................................................................................. 77

Figure 4-4: Dimension drawing of IC 7404 ........................................................................ 79

Figure 4-5: Bread board testing of the circuit ..................................................................... 81

Figure 4-6: PCB implementation ...................................................................................... 82

Figure 4-7: Final circuit implemented on PCB .................................................................... 82

Figure 4-8: Connections with the ADuC842 microcontroller ................................................. 83

Figure 4-9: PICASO-GFX2 microcontroller from 4D labs ..................................................... 83

Figure 4-10: PICASO-GFX2 µcontroller which is in-built on back panel of OLED module ...... 85

Figure 4-11: PICASO-GFX2 internal block diagram ............................................................ 86

Figure 4-12: Pin diagram of PICASO-GFX2 microcontroller ................................................. 86

Figure 4-13: PICASO-GFX2 internal memory organization ................................................... 90

Figure 4-14: Waveform for the heart beat .......................................................................... 92

Figure 5-1: Configuring the Keil compiler and assembler .................................................... 94

Figure 5-2: Compiler settings .......................................................................................... 94

Figure 5-3: Project menu showing 'New Project' command option .......................................... 95

Figure 5-4: Project menu showing 'Open Project' command option ......................................... 96

Figure 5-5: Project menu showing 'Save Project' command option .......................................... 96

Figure 5-6: Project menu showing 'Close Project' command option ......................................... 97

Figure 5-7: Project menu showing 'Remove' command option ................................................ 98

Figure 5-8: Project menu showing 'Project Settings' command option ..................................... 99

Figure 5-9: Project menu showing 'Insert' command option ................................................. 100

Figure 5-10: Dialog box showing 'Add Files to Folder' command option ............................... 100

Figure 5-11: Dialog box showing 'Compile' command option .............................................. 102

Figure 5-12: O/P window showing compilation statistics & error summary on Build tab ......... 103

Figure 5-13: O/P window showing error message & line marked with blue dot ...................... 103

Figure 5-14: Dialog box showing 'Build' command option..................................................... 104

Figure 5-15: Dialog box showing 'Rebuild All' command option........................................... 104

Figure 5-16: Flowchart for the calculation of blood oxygen saturation and heart rate.............. 106

Figure 5-17: Workspace for writing the program ...................................................................... 108

Figure 5-18: Program showing zero errors and zero warnings................................................... 108

Figure 5-19: Compilation of the program .................................................................................. 109

Figure 5-20: Running of the program ........................................................................................ 109

Figure 5-21: Results of the program .......................................................................................... 110

Figure 6-1: Display of heart rate and blood oxygen saturation along with the waveform ......... 111

# LIST OF TABLES

Table 2-1: Normal pulse rates in beats per minute (BPM) ............................ 14

Table 2-2: SpO2 reading and its interpretation........................................... 17

Table 3-1: Source Impedance and DC Accuracy....................................... 35

Table 3-2: Some Single-Supply Op-Amps ................................................ 36

Table 3-3: ADCCON1 SFR bit designations.............................................. 39

Table 3-4: ADCCON2 SFR bit designations.............................................. 40

Table 3-5: ADCCON3 SFR bit designations.............................................. 41

Table 3-6: Pin description of ADuC842 microcontroller ............................ 47

Table 3-7: PSW SFR bit designations........................................................ 56

Table 3-8: PCON SFR bit designations ..................................................... 57

Table 3-9: TIMECON SFR bit designations............................................... 59

Table 3-10: TMOD SFR bit designations ................................................. 62

Table 3-11: TCON SFR bit designations.................................................... 63

Table 3-12: T2CON SFR bit designations.................................................. 64

Table 3-13: Pin description of power, serial and micro-USB interface........................ 71

Table 3-14: Pin description of Expansion Port J1...................................... 72

Table 3-15: Pin description of Expansion Port J2....................................... 73

Table 4-1: Pin description of IC 7404........................................................ 78

Table 4-2: Technical data of IC 7404 ....................................................... 80

Table 4-3: Pin description of PICASO-GFX2 microcontroller ..................... 87

Table 4-4: Operating conditions of PICASO-GFX2 microcontroller........................... 91

Table 4-5: Global characteristics based on operating conditions of PICASO-GFX2 ................. 91

Table 5-1: SpO2 Interpretation ............................................................... 105

Table 5-2: ASCII code table .................................................................... 107

# CHAPTER 1

# INTRODUCTION

Instrumentation devices have become a staple in the medical world, both in research laboratories and clinical settings. They have become so relied upon, that physicians, nurses, occupational therapists, physical therapists and other health care professionals cannot carry on their work suitably without them. From ECG's to EMG's, EEG's and pulse oximeters, the list continues. These devices have not only revolutionized the field of medicine with their rich technology, but have also increased the quality of medical care that patients receive. The purpose of these devices is to provide vital information, usually on a continuous basis, to clinicians and other health care professionals in order so that they may be able to set a course of action, make decisions or plan the next steps whether these are on treatment options, monitoring options etc. However, like any system or device, there is always room for improvements and advancements.

For many years, electronics have been built into medical instruments but mainly focused on large expensive equipments such as CAT and MRI scanners used in clinics and hospitals. Recently, there has been an increased desire for medical instruments that can be used at home as well as in clinics. The advantage of using medical instruments at home allows the patient to monitor their health constantly without the need to visit the clinic for a check up. This is especially useful for the elderly where it can be difficult for them to make visits to the clinics, so the development of small, reliable, accurate medical devices is now accelerating in the current market.

These small instruments have to be energy efficient and patient friendly in both physically and technically. Creating such devices can be challenging, especially now that the state of the art medical products are non-invasive so it does not intimidate the patient when using the device.

## 1.1 History of Pulse Oximeter

In times of doctor shortages and ever increasing wait times in hospitals and clinics, we hope to implement a system that will be able to decrease both the wait times of a patient and the work load of a physician. The pulse oximeter is a diagnostic tool used world-wide in medicine to monitor the oxygenation of a patient's blood level as well as the heart rate. Its history goes back to the 1850s. Credit for its development involved much research, studies and experiments by early scientists. The pulse oximeter has made a significant impact on the medical field and has propelled the advancement of patient care particularly in the areas of anesthesia and critical care.

1

The following outlines the development of this important device:

1850s- Russian physiologist I.M. Sechenov developed a vacuum blood pump which was laterly used for research purpose.

1864- Sir George Gabriel Stokes, 1st Baronet, an Irish physicist and mathematician, discovered the respiratory function of haemoglobin.

1876- Karl von Vierordt, a German physician who developed techniques and tools for the monitoring of blood circulation, used a light source to distinguish fully saturated blood from that which is not.

1898- English physiologist Halden brought forward the principle of chemical expulsion of oxygen from its complexes with haemoglobin. J.Barcoft used this principle for the examination of gas composition of blood.

1900- R. Vierordt demonstrated that application of Hufner tourniquet caused a decrease in the intensity of red light passed through a human hand.

1922- American biochemist D. van Slyke combined the vacuum and chemical principles of gas expulsion from blood and used them in his manometric appratus.

1932- German physiologist L. Nicolai optically recorded the in vivo oxygen consumption of a hand after circulatory occlusion in Gottingen, Germany.

1935- Matthes developed the first 2-wavelength ear $O_2$ saturation meter with red and green filters, later switched to red and infrared filters. This was the first device to measure $O_2$ saturation.

1949- Wood added a pressure capsule to squeeze blood out of ear to obtain zero setting in an effort to obtain absolute $O_2$ saturation value when blood was readmitted. The concept is similar to today's conventional pulse oximetry but was hard to implement because of unstable photocells and light sources. This method is not used clinically.

1964- Shaw assembled the first absolute reading ear oximeter by using eight wavelengths of light. Commercialized by Hewlett Packard, its use was limited to pulmonary functions and sleep laboratories due to cost and size.

1974- Pulse oximetry was developed by Takuo Aoyagi and Michio Kishi, bioengineers, at Nihon Kohden using the ratio of red to infrared light absorption of pulsating components at the measuring site.

1975- Susumu Nakajima, a surgeon, and his associates first tested the device on patients.

1979- Biox was founded.

1981- Biox introduced the first pulse oximeter to commercial distribution.

1982- Biox initially focused on respiratory care, but when the company discovered that their pulse oximeters were being used in operating rooms to monitor oxygen levels, Biox expanded its marketing resources to focus on operating rooms.

1983- A competitor, Nellcor, began to compete with Biox for the US operating room market.

1987- The standard of care for the administration of a general anesthetic in the US included pulse oximetry. From the operating room, the use of pulse oximetry rapidly spread throughout the hospital, first to the recovery room, and then into the various intensive care units (ICUs). Pulse oximetry was of particular value in the neonatal unit where the patients do not thrive with inadequate oxygenation, but also can be blinded with too much oxygen. Furthermore, obtaining an arterial blood gas from a neonatal patient is extremely difficult.

1995- Masimo introduced Signal Extraction Technology (SET) that could measure accurately during patient motion and low perfusion. Some have termed newer generation pulse oximetry technologies as High Resolution Pulse Oximetry (HRPO). One area of particular interest is the use of pulse oximetry in conducting portable and in-home sleep apnea screening and testing.

2009- The world's first Bluetooth-enabled fingertip pulse oximeter was introduced by Nonin Medical, enabling clinicians to remotely monitor patients' pulses and oxygen saturation levels. It also allows patients to monitor their own health through online patient health records and home telemedicine system.

Today there are many manufacturers of pulse oximeters. All offer a variety of different oximeter boxes with $SpO_2$ and pulse rate readings, waveform displays, alarms etc. These days "new-generation" pulse oximeters have been introduced that have elevated the accuracy of pulse oximeter readings significantly. Now-a-days, pulse oximeters have become affordable and widely available for home use.

## 1.2 General Concepts

With the advancement of science, new and sophisticated technologies have evolved which play a critical role in making medical sciences reliable and accurate. Bio-medical instrumentation is an ever evolving field with a scope of development of high precision equipments that will help save lives of many people. Health has been one of the most significant considerations in the fortification of human civilization. A lot of resource has been used in health sector to aid mankind fight diseases of all kinds. Different researches are ongoing for development of sophisticated high precision devices which make medical services reliable and accurate.

Introduced in the early 1980s, pulse oximeter is a medical device that indirectly monitors the oxygen saturation of a patient's blood (as opposed to measuring oxygen saturation directly through a blood sample) and changes in blood volume in the skin. A pulse oximeter also measures the heart rate of a patient. It is often attached to a medical monitor so staff can see a patient's oxygenation at all times. A pulse oximeter is a particularly convenient, noninvasive measurement instrument whereas with arterial blood gas sampling, the process becomes invasive, difficult, painful, expensive and potentially risky. Pulse oximeter is one such theory employed to develop a sensor capable of monitoring saturation of oxygen in blood cells as well as heart rate. The principle advantage of optical sensors for medical applications is their intrinsic safety as there is no electrical contact between the patient and the equipment. Hemoglobin, the colored substance in blood is a carrier of oxygen. It combines with oxygen to form oxidized hemoglobin. The absorption of light by hemoglobin varies with oxygen concentration.

For patients at risk of respiratory failure, it is important to monitor the efficiency of gas exchange in the lungs, i.e. how well the arterial blood is oxygenated (as opposed to whether or not air is going in and out of the lungs). Preferably, such information should be available to the doctors in a continuous basis (rather than every few hours). These requirements can be met non-invasively with the technology of pulse oximetry. The key features of pulse oximeter are:

1) Small enough to be wearable
2) Less expensive compared to the old health monitoring devices
3) Simple to use , needs no user calibration
4) Accurate enough for clinical use

For these reasons, in almost every hospital, critical care units and surgical theaters; the pulse oximeter is acknowledged as **standard monitoring device**.

Figure 1-1 shows the benchtop pulse oximeter displaying the waveform as well as showing the values for blood oxygen saturation and heart rate.

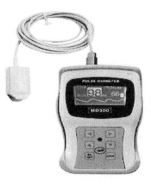

Figure 1-1: Benchtop Pulse Oximeter

## 1.3 Advantages and Disadvantages of Pulse Oximeter

### 1.3.1 Advantages of Pulse Oximeter

1) Pulse oximetry system is advantageous for health status of elderly for a number of reasons. The failure of oxygen delivery to the heart or brain can result to death. Monitoring of oxygen delivery is vital and pulse oximetry measures arterial oxygen saturation ($SpO_2$).

2) Another benefit of pulse oximetry is the capability to measure other critical physiological information from a single compact sensor like the heart rate.

3) Pulse oximetry became accepted for a noninvasive method and it can give immediate data of the arterial oxygen saturation in the patient's blood.

4) Because of their simplicity and speed, pulse oximeters are of critical importance in emergency medicine and are also very useful for patients with respiratory or cardiac problems.

5) Portable, battery operated pulse oximeters are useful for pilots operating in a non-pressurized aircraft above 10,000 feet (12,500 feet in the US) where supplemental oxygen is required.

6) Portable pulse oximeters are also useful for mountain climbers and athletes whose oxygen levels may decrease at high altitudes or with exercise.

7) Pulse oximetry measures a patient's arterial saturation of oxygen ($SpO_2$) in seconds; measurements obtained from Arterial Blood Gas (ABG) can take several minutes to acquire and are usually drawn by a respiratory therapist, transported to the blood gas lab, measured

5

and finally reported back to the doctor. The information from an ABG analysis provides for more information than just the oxygen saturation level and is essential in emergencies.

8) Accurate and easy to use.
9) Non-invasive and reliable

## 1.3.2 Disadvantages of Pulse Oximeter

1) Critically ill Patients: It may be less effective in very sick patients, because tissue perfusion may be poor and thus the oximeter probe may not detect a pulsatile signal.

2) Waveform Presence: If there is no waveform visible on a pulse oximeter, any percentage saturation values obtained are meaningless.

3) Inaccuracies:

 i. Bright overhead lighting, shivering and motion artifact may give pulsatile waveforms and saturation values when there is no pulse.

 ii. Abnormal haemoglobins such as methaemoglobinaemia, for example following overdose of prilocaine, cause readings to tend towards 85%.

 iii. Dyes and pigments, including nail varnish, may give artificially low values.

 iv. Vasoconstriction and hypothermia cause reduced tissue perfusion and failure to register a signal.

 v. Carboxyhaemoglobin, caused by carbon monoxide poisoning, causes saturation values to tend towards 100%. A pulse oximeter is extremely misleading in cases of carbon monoxide poisoning for this reason and should not be used. Co-oximetry is the only available method of estimating the severity of carbon monoxide poisoning.

 vi. Rare cardiac valvular defects such as tricuspid regurgitation cause venous pulsation and therefore venous oxygen saturation is recorded by the oximeter.

 vii. Oxygen saturation values less than 70% are inaccurate as there are no control values to compare them to.

 viii. Cardiac arrhythmias may interfere with the oximeter picking up the pulsatile signal properly and with calculation of the pulse rate.

4) Response delay due to signal averaging: This means that there is a delay after the actual oxygen saturation starts to drop because the signal is averaged out over 5 to 20 seconds.

5) Not a monitor of ventilation: A recent case report highlighted the false sense of security provided by pulse oximetry. An elderly woman postoperatively in the recovery room was receiving oxygen by face mask. She became increasingly drowsy, despite having an oxygen saturation of 96%. The reason was that her respiratory rate and minute volume were low due to residual

neuromuscular block and sedation, yet she was receiving high concentrations of inspired oxygen, so her oxygen saturation was maintained. She ended up with an arterial carbon dioxide concentration of 280 mmHg (normal 40 mmHg) and was ventilated for 24 hours on intensive care. Thus oximetry gives a good estimation of adequate oxygenation, but no direct information about ventilation, particularly as in this case, when supplemental oxygen is being administered.

6) Lag monitor: This means that the partial pressure of oxygen can have fallen a great deal before the oxygen saturation starts to fall. If a healthy adult patient is given 100% oxygen to breathe for a few minutes and then ventilation ceases for any reason, several minutes may elapse before the oxygen saturation starts to fall. A pulse oximeter in these circumstances warns of a potentially fatal complication several minutes after it has happened. The pulse oximeter has been described as "a sentry standing at the edge of the cliff of desaturation" because of this fact.

7) Patient safety: There has been one or two case reports of skin burns or pressure damage under the probe because some early probes had a heater unit to ensure adequate skin perfusion. The probe should be correctly sized, and should not exert excessive pressure. Special probes are now available for paediatric use.

## 1.4 Applications of Pulse Oximeter

1) Monitoring patients during anesthesia, intensive care, emergency departments, general wards, or those with conditions such as asthma.

2) Pulse oximeters detect the presence of cyanosis more reliably than even the best doctors when using their clinical judgment.

3) In aircraft, helicopters or ambulances, the audible tone and alarms may not be heard; but if a waveform can be seen together with an acceptable oxygen saturation, this gives a global indication of a patient's cardio-respiratory status.

4) To assess the viability of limbs after plastic and orthopedic surgery and, for example, following vascular grafting, or where there is soft tissue swelling; as a pulse oximeter requires a pulsatile signal under the sensor, it can detect whether a limb is getting a blood supply or not.

5) To limit oxygen toxicity in premature neonates supplemental oxygen can be tapered to maintain an oxygen saturation of 90% - thus avoiding the damage to the lungs and retinas of neonates.

6) During thoracic anesthesia - when one lung is being collapsed down - to determine whether oxygenation via the remaining lung is adequate or whether increased concentrations of oxygen must be given.

7) Clinical applications of pulse oximetry include anesthesia, PACU, intensive care units (ICUs), neonatal (intensive care, newborn nursery and delivery suites), transport (within the hospital,

external and ambulance/air transport), diagnostic lab, subacute care centers, home care patients, other patient conditions (blood flow assessment, cardiopulmonary arrest, asthma and seizures) etc.

8) As a means of reducing the frequency of blood gas analysis in intensive care patients- especially in pediatric practice where vascular (arterial) access may be more difficult.

9) Fetal oximetry- a developing technique that uses reflectance oximetry, using LEDs of 735nm and 900nm. The probe is placed over the temple or cheek of the fetus, and needs to be sterile and sterilisable. They are difficult to secure and the readings are variable, for physiological and technical reasons. Hence the trend is more useful than the absolute value.

## 1.5 Comparison of pulse oximeter with Arterial Blood Gas (ABG) sampling

1) Pulse oximeter measures blood oxygen saturation in seconds whereas Arterial Blood Gas sampling technique can take several minutes.

2) Pulse oximeter is a non-invasive method of measurement whereas Arterial Blood Gas is an invasive method of measurement.

3) The information from an ABG analysis provides far more information than just the oxygen saturation level and is essential in emergencies whereas pulse oximeter only tells the blood oxygen saturation level.

4) Pulse oximeter can also tell the heart rate of a patient whereas ABG analysis can not tell the heart rate.

5) Pulse oximeter is an easy way of measurement than ABG analysis which is difficult and painful.

6) Pulse oximeter is more accurate and reliable method of measurement than ABG analysis.

## 1.6 Problem Statement

1) To calculate blood oxygen saturation and heart rate in digital values and display the same on the OLED.

2) To display the waveform for heart beat.

# CHAPTER 2

# DETAILS OF TECHNOLOGY USED

## 2.1 Principle of pulse oximetry technology

The principle of pulse oximetry is based on the red and infrared light absorption characteristics of oxygenated and deoxygenated hemoglobin. Oxygenated hemoglobin absorbs more infrared light and allows more red light to pass through. Deoxygenated (or reduced) hemoglobin absorbs more red light and allows more infrared light to pass through. Red light is in the 600-750 nm wavelength light band. Infrared light is in the 850-1000 nm wavelength light band. Figure 2-1 shows the absorption relationship of oxygen levels in the blood for the red and infrared wavelengths.

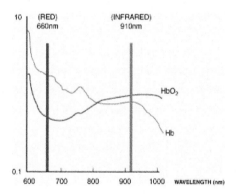

Figure 2-1: Absorption relationship of oxygen levels in the blood for the red and IR wavelengths

Pulse oximetry uses a light emitter with red and infrared LEDs that shines through a reasonably translucent site with good blood flow. Typical adult/pediatric sites are the finger, toe, pinna (top) or lobe of the ear. Infant sites are the foot or palm of the hand and the big toe or thumb. Opposite the emitter is a photo detector that receives the light that passes through the measuring site.

At the measuring site there are constant light absorbers that are always present. They are skin, tissue, venous blood, and the arterial blood. However, with each heart beat the heart contracts and there is a surge of arterial blood, which momentarily increases arterial blood volume across the measuring site. This results in more light absorption during the surge. If light signals received at the photodetector are looked at 'as a waveform', there should be peaks with each heartbeat and troughs between heartbeats. If the light absorption at the trough (which should include all the constant

9

absorbers) is subtracted from the light absorption at the peak, then in theory, the resultants are the absorption characteristics due to the added volume of blood only; which is arterial. Since peaks occur with each heartbeat or pulse, the term "pulse oximetry" was coined. This solved many problems inherent to oximetry measurements in the past and is the method used today in conventional pulse oximetry.

Still, conventional pulse oximetry accuracy suffered greatly during motion and low perfusion and made it difficult to depend on when making medical decisions. Arterial blood gas tests have been and continue to be commonly used to supplement or validate pulse oximeter readings. The advent of "Next Generation" pulse oximetry technology has demonstrated significant improvement in the ability to read through motion and low perfusion; thus making pulse oximetry more dependable to base medical decisions on.

The main operation of a pulse oximeter is the determination of person's functional oxygen saturation. Arterial oxygen saturation, or $SaO_2$, is the percentage of functional arterial hemoglobin that is oxygenated. Functional hemoglobin is a type of hemoglobin that is able to bind with oxygen. Non-functional hemoglobin cannot bind with oxygen. An example of non-functional hemoglobin is carboxyhemoglobin (COHb), which binds easily with carbon monoxide. When functional hemoglobin binds with four oxygen molecules, it is considered oxygenated hemoglobin ($HbO_2$). When it is carrying less than four oxygen molecules, it is considered reduced hemoglobin. Functional oxygen saturation measured with a pulse oximeter is often called $SpO_2$ because it is estimation based peripheral measurements and an assumption that only $HbO_2$ and Hb are present in the blood. The presence of non-functional hemoglobin such as COHb can cause erroneous measurements. Therefore, $SpO_2$ is a different measurement than $SaO_2$. $SpO_2$, stands for "Saturation of Peripheral Oxygen". It refers to the concentration of oxygen in a patient's periphery; in this case, it is usually measured in a fingertip or earlobe.

There are two methods of sending light through the measuring site:

1. Transmission
2. Reflectance

In the transmission method, the emitter and photo detector are opposite of each other with the measuring site in-between. The light can then pass through the site. In the reflectance method, the emitter and photo detector are next to each other on top of the measuring site. The light bounces from the emitter to the detector across the site. The transmission method is the most common type used.

Figure 2-2 shows the transmission and reflectance type of sensor.

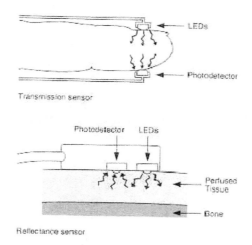

Transmission sensor

Reflectance sensor

Figure 2-2: Transmission and reflectance type of sensor

More elaborated figures of transmission and reflectance pulse oximetry are shown in figure 2-3.

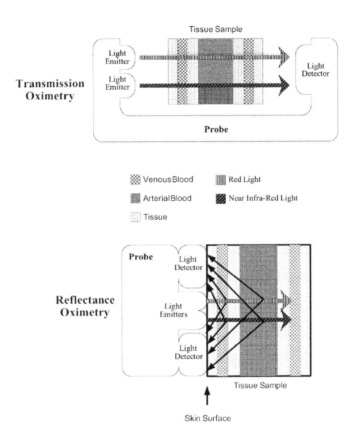

Figure 2-3: The principle difference between transmission and reflectance pulse oximetry

After the transmitted red (R) and infrared (IR) signals pass through the measuring site and are received at the photo detector, the R/IR ratio is calculated. The R/IR is compared to a "look-up" table (made up of empirical formulas) that converts the ratio to a $SpO_2$ value. Typically an R/IR ratio of 0.5 equates to approximately 100% $SpO_2$, a ratio of 1.0 equates to approximately 82% $SpO_2$, while a ratio of 2.0 equates to 0% $SpO_2$. Most manufacturers have their own look-up tables based on calibration curves derived from healthy subjects at various $SpO_2$ levels.

Because the flow of blood is pulsatile in nature, the transmitted light changes with time. A normal finger has light absorbed from bloodless tissue, venous blood, and arterial blood. The volume of arterial blood changes with pulse, so the absorption of light also changes. The light detector will therefore see a large DC signal representing the residual arterial blood, venous blood, and bloodless tissue. A small portion of the detected signal (~1%), will be an AC signal representing the arterial pulse. Because this is the only AC signal, the arterial portion of the signal can be differentiated. This AC signal is separated with simple filtering and an RMS value can be calculated.

Figure 2-4 and 2-5 shows the light absorption by tissues.

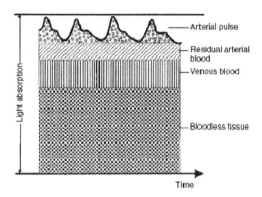

Figure 2-4: Light absorption by tissue type

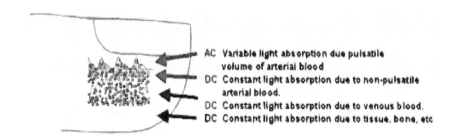

Figure 2-5: Light absorption by tissues

## 2.2 Working of pulse oximeter

A pulse oximeter is a blood oxygen monitor using red and infra red light to measure the percentage of oxygen that is saturated within an individual's blood. Red blood cells are red because they contain protein chemical called hemoglobin. Hemoglobin contains iron, which is utilized for transporting oxygen and carbon dioxide. When blood passes through the lungs, oxygen molecules attach to the hemoglobin making them oxyhemoglobin. Hemoglobin releases oxygen while blood passes through the body's tissue becoming deoxyhemoglobin. The empty hemoglobin molecules then bond with the tissue's carbon dioxide or other waste gases, transporting it away; during this process, the color of blood changes. The pulse oximeter detects these changes through two light emitting diodes and two sensors on the opposite side of the finger. One diode sends out red light while the other sends out infrared light. Oxygenated blood absorbs light at 660nm (red light), whereas deoxygenated blood absorbs light preferentially at 940nm (infra-red). The relative absorption of light by oxyhemoglobin (HbO) and deoxyhemoglobin is processed by the device and an oxygen saturation level is reported. The body naturally delivers oxygenated blood in arteries and returns deoxygenated blood in veins. To ensure only the oxygenated blood levels are measured, pulse oximeters are programmed to detect pulsatile blood, which only occurs in veins, thus preventing oximeters from reading blood that has been deoxygenated.

**2.3 Design Parameters**

    **2.3.1**    **Heart Rate:** Heart rate is defined as the number of beats in one minute.

    i.    For a new born: A newborn's heart rate is typically around 120 beats per minute (bpm)

    ii.    For an adult: A heart rate in the vicinity of 70 beats per minute (bpm) is considered normal.

    iii.    When a person enters his golden years, the heart rate slows to approximately 50 bpm.

When exercising, the heart rate may double. Accounting for all of this data, to say, 50 to 200 bpm are considered good readings for the heart rate.

Table 2-1 shows the normal pulse rates in beats per minute (BPM).

Table 2-1: Normal pulse rates in beats per minute (BPM)

| newborn | 1 - 12 months | 1 - 2 years | 2 - 6 years | 6 - 12 years | 12 years - adults |
|---------|---------------|-------------|-------------|--------------|-------------------|
| 120 - 160 | 80 - 140 | 80 - 130 | 75 - 120 | 75 - 110 | 60 - 100 |

Pulse is the pressure wave of blood that is generated when our heart muscles contract. It reflects the rhythm, rate and strength of our heart's contractions. We can feel our pulse anywhere that an artery (a blood vessel that carries blood away from the heart) crosses over a bone and is close to the skin's surface. There are a few things to keep in mind when you're finding your pulse. The first is: don't use your thumb. Your thumb has a pulse of its own, which can trick you into thinking you've found a pulse point when you haven't. This is mainly a concern when you're feeling for someone else's pulse-you don't want to confuse your pulse with theirs. To avoid this problem, always use your forefinger and middle finger to locate a pulse.

In medicine, a person's pulse represents the tactile arterial palpation of the heartbeat. The pulse may be palpated in any place that allows an artery to be compressed against a bone, such as at the neck (carotid artery), at the wrist (radial artery), behind the knee (popliteal artery), on the inside of the elbow (brachial artery), and near the ankle joint (posterior tibial artery). The pulse rate can also be measured by measuring the heart beat directly (auscultation), usually using a stethoscope.

The common pulse sites are:

a) Upper limb:
  i. Axillary pulse: Located inferiorly of the lateral wall of the axilla.
  ii. Brachial pulse: Located on the inside of the upper arm near the elbow, frequently used in place of carotid pulse in infants (brachial artery).
  iii. Radial pulse: Located on the lateral of the wrist (radial artery). It can also be found in the anatomical snuff box.
  iv. Ulnar pulse: Located on the medial of the wrist (ulnar artery).

Figure 2-6 shows the front of right upper extremity.

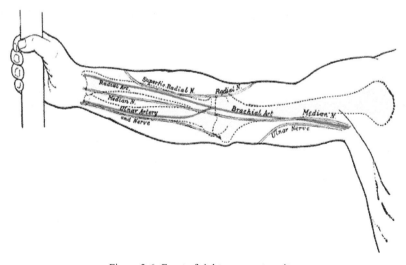

Figure 2-6: Front of right upper extremity

b) Lower limb:

   i.   Femoral pulse: located in the inner thigh, at the mid-inguinal point, halfway between the pubic symphysis and anterior superior iliac spine (femoral artery).

   ii.  Popliteal pulse: Above the knee in the popliteal fossa, found by holding the bent knee. The patient bends the knee at approximately 124°, and the physician holds it in both hands to find the popliteal artery in the pit behind the knee (Popliteal artery).

   iii. Dorsalis pedis pulse: Located on top of the foot, immediately lateral to the extensor of hallucis longus (dorsalis pedis artery).

   iv.  Tibialis posterior pulse: Located on the medial side of the ankle, 2 cm inferior and 2 cm posterior to the medial malleolus (posterior tibial artery). It is easily palpable over Pimenta's Point.

c) Head/Neck:

   i.   Carotid pulse: Located in the neck (carotid artery). The carotid artery should be palpated gently and while the patient is sitting or lying down. Stimulating its baroreceptors with low palpitation can provoke severe bradycardia or even stop the heart in some sensitive persons. Also, a person's two carotid arteries should not be palpated at the same time. Doing so may limit the flow of blood to the head, possibly leading to fainting or brain ischemia. It can be felt between the anterior border of the sternocleidomastoid muscle, above the hyoid bone and lateral to the thyroid cartilage.

   ii.  Facial pulse: Located on the mandible (lower jawbone) on a line with the corners of the mouth (facial artery).

   iii. Temporal pulse: Located on the temple directly in front of the ear (superficial temporal artery).

Figure 2-7 shows the arteries of the neck.

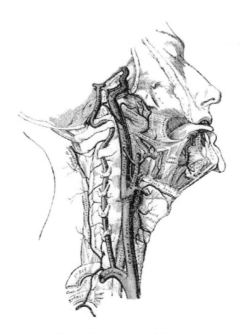

Figure 2-7: Arteries of the neck

**2.3.2 Blood oxygen saturation:** For patients at risk of respiratory failure, it is important to monitor the blood oxygen content of such individuals to ensure proper perfusion of blood in their system. This is important as our body tissues require a constant supply of oxygen, and without it can only survive for a very short period of time. Preferably this information should be received on a continuous basis.

Blood oxygen saturation is interpreted with the following table:

Table 2-2: $SpO_2$ reading and its interpretation

| $SpO_2$ Reading (%) | Interpretation |
|---|---|
| 95-100 | Normal |
| 91-94 | Mild Hypoxemia |
| 86-90 | Moderate Hypoxemia |
| <85 | Severe Hypoxemia |

where Hypoxemia is defined as decreased partial pressure of oxygen in blood and low oxygen availability to the body or an individual tissue or organ.

## 2.4 Different types of pulse oximeters

The different types of pulse oximeters and their detailed description are described below:

1) Fingertip pulse oximeter: A fingertip pulse oximeter, which clips onto a patient's finger, has a tiny computer and screen. The clip emits light from one side and measures the light on the other. The computer measures the light over several pulses and gives readout of the patient's blood-oxygen level. Often used at home, fingertip oximeters are the easiest type to operate. However, if the patient's hand is damaged or his blood flow is slow because of a clot or injury, the fingertip oximeter may not give an accurate reading or it may not work at all.

Figure 2-8: Finger tip pulse oximeter

2) Handheld pulse oximeter: The handheld pulse oximeter, used in almost every hospital, is similar to the fingertip device but has a more versatile clip that connects with a cord to a computer. Like the fingertip oximeter, it uses light to measure the blood's hemoglobin. It clips to a patient's fingertip or earlobe. This is important in cases where a patient's extremities are compromised and may not have full blood flow. Since it is so close to the brain, it is rare for an earlobe to have compromised blood flow. In an emergency, health-care professionals clip a handheld oximeter to a patient's toe.

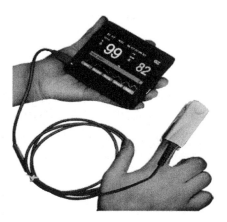

Figure 2-9: Handheld pulse oximeter

3) Fetal pulse oximeter: It is vital to know the fetal oxygen level, especially during labor. But it is difficult, if not impossible, to place a traditional probe on a baby before birth. A fetal pulse oximeter solves this problem. Developed in the 1990's this device has a probe that doctors can insert into the birth canal and place on the baby's skull. The sensor shoots light across the probe and measures the hemoglobin level in the baby's scalp. The sensor attaches to a cord leading to a computer. One of the benefits of this oximeter is that the probe can also measure the baby's heart beats.

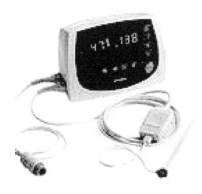

Figure 2-10: Fetal pulse oximeter

## 2.5 Difference between blood and hemoglobin

It is obviously known that our body consists of blood. Blood serves many functions including transportation of $O_2$, $CO_2$, nutrients, heat and hormones to the different tissues of the body, regulation of various aspects of the body including temperature, pH, and water content of the cells and protection from diseases and loss of blood. Blood consists of 3 different components:

1) Plasma, which is the liquid or "watery" component of the blood. This constitutes 55%.

2) Erythrocytes, better known as red blood cells which constitute 45%, and

3) Leukocytes, better known as white blood cells, which constitute less than 1%.

Each of these components is required to perform its function. Red blood cells and white blood cells are, in essence, completely different. While both are necessary for the body's proper functioning, they each have singular roles.

Red blood cells, also called erythrocytes, are responsible for the characteristic color of our blood. They are responsible for picking up carbon dioxide from our blood and for transporting oxygen. Red blood cells are shaped like a biconcave disc as shown in figure 2-11. White blood cells or leukocytes, on the other hand, are primarily responsible for fighting foreign organisms that enter the body. This includes everything from bacterial and parasitic infections to allergic response. T-cells, a form of white blood cells, are the ones that stop functioning properly in the presence of an HIV infection. An overproduction of white blood cells can lead to leukemia.

There are approximately 5 million red blood cells in every cubic millimeter of blood and only 3,000 - 7,000 white blood cells in the same amount of blood. Red blood cells have an average lifespan of 120 days, while white cells live a maximum of four days.

Red blood cells have a circular shape that resembles a shallow bowl, but they can change shape without breaking to squeeze through smaller spaces if necessary. White blood cells have different shapes, depending on their function. While they can multiply easily, they don't change shape.

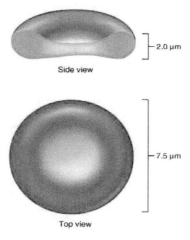

2.0 μm

Side view

7.5 μm

Top view

Figure 2-11: Red blood cell structure

Hemoglobin (protein-iron complex) is the essential component of red blood cells, which can hold oxygen so the cells can then transport around the body to the peripheral tissues. Hemoglobin also removes a limited amount of carbon dioxide from the peripheral tissues. This process is what gives the body energy, which explains why people who suffer from anemia— low count red blood cells— often, feel tired and sleepy. A high count of red blood cells is rare, but it can happen. Causes include kidney disease, dehydration, anabolic steroid use, and pulmonary fibrosis. People suffering from a high count of red blood cells usually have impaired circulation, and are at a high risk for heart disease.

Hemoglobin consists of four different protein structures (2 alpha chains and 2 beta chains) and each protein consists of a heme molecule that contains a centrally located iron. The heme molecule is actually responsible for binding with oxygen, and therefore each haemoglobin molecule can bind up to 4 oxygen molecules. A typical red blood cell consists of millions of hemoglobin molecules! Hemoglobin protein and iron-containing heme group is depicted in the figure 2-12, showing the details of the protein structure as well as the heme group which binds to oxygen.

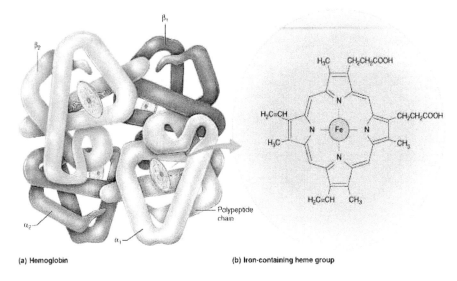

(a) Hemoglobin          (b) Iron-containing heme group

Figure 2-12: Hemoglobin protein and iron-containing heme group

Once air is breathed into the lungs, blood that is pumped into the pulmonary system exchanges $CO_2$ and takes up (binds to) $O_2$. This is known as gas exchange in the blood. This occurs due to a partial pressure gradient that exists between the blood and inhaled air. Blood coming into the lungs has a lower partial pressure of $O_2$ and a higher partial pressure of $CO_2$. Due to the laws of gases, a gas particle will diffuse (move) from an area of higher pressure to lower pressure. Therefore this partial pressure gradient that exists facilitates the movement of oxygen to the red blood cells and carbon dioxide back out into the environment.

Hemoglobin has a property that when bound to oxygen it causes the hemoglobin to bind more easily to more oxygen. This gives a characteristic oxygen-hemoglobin dissociation curve that is well known to physiologists and is presented below in figure 2-13.

# Gas Transport in Blood

## Oxygen-Hemoglobin Dissociation Curve

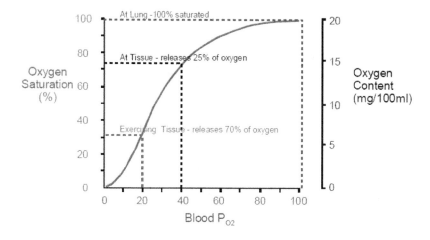

Figure 2-13: Oxygen-Hemoglobin dissociation curve

From this figure we can conclude that arterial blood must be close to 100% saturated when delivering blood to the various body tissues. We can also see from this figure that once oxygen is also released (or taken up), this facilitates more oxygen to be released and therefore gives a characteristic "S" shaped curve.

# CHAPTER 3

# WORK DONE

## 3.1 Block Diagram

The block diagram of pulse oximeter as shown in figure 3-1 consists of the following stages:

1. PPG probe
2. Switching and LED driver circuit
3. I/V converter
4. Amplifier stage
5. ADC stage
6. Microcontroller
7. LCD/OLED display

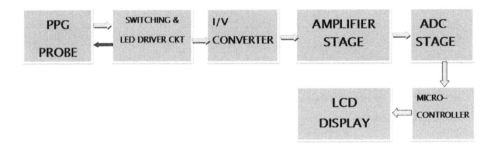

Figure 3-1: Block diagram of pulse oximeter

### 3.1.1 PPG (Photoplethysmograph) Probe

**3.1.1.1 Plethysmograph:** A plethysmograph is an instrument for measuring changes in volume within an organ or whole body (usually resulting from fluctuations in the amount of blood or air it contains). We can also say that plethysmograph is an instrument that measures variations in the size of an organ or body part on the basis of the amount of blood passing through or present in the part. Organs studied are lungs, limbs etc. Each organ employs different type of sensor and for different age groups too.

**3.1.1.2 PPG (Photoplethysmograph):** A photoplethysmograph (PPG) is an optically obtained plethysmograph, a volumetric measurement of an organ. A PPG is often obtained by using a pulse oximeter which illuminates the skin and measures changes in light absorption.

With each cardiac cycle, the heart pumps blood to the periphery. Even though this pressure pulse is somewhat damped by the time it reaches the skin, it is enough to distend the arteries and arterioles in the subcutaneous tissue.

The change in volume caused by the pressure pulse is detected by illuminating the skin with the light from a light-emitting diode (LED) and then measuring the amount of light either transmitted or reflected to a photodiode. Each cardiac cycle appears as a peak as shown in figure 3-2.

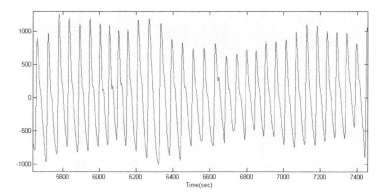

Figure 3-2: Representative PPG taken from an ear pulse oximeter;

Variations in amplitude are from Respiratory Induced Variation

Because blood flow to the skin can be modulated by multiple other physiological systems, the PPG can also be used to monitor breathing, hypovolemia, and other circulatory conditions.

Additionally, the shape of the PPG waveform differs from subject to subject, and varies with the location and manner in which the pulse oximeter is attached.

Figure 3-3 shows the different layers of human skin.

25

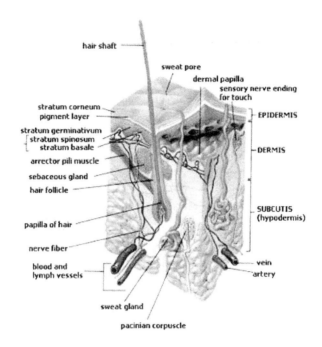

Figure 3-3: Layers of human skin

While pulse oximeters are a commonly used medical device, the PPG derived from them is rarely displayed and is nominally only processed to determine heart rate. PPGs can be obtained from transmissive absorption (as at the finger tip) or reflective (as on the forehead). In outpatient settings, pulse oximeters are commonly worn on the finger. However, in cases of shock, hypothermia, etc. blood flow to the periphery can be reduced, resulting in a PPG without a discernible cardiac pulse. In this case, a PPG can be obtained from a pulse oximeter on the head, with the most common sites being the ear, nasal septum, and forehead.

Uses of PPG are monitoring heart rate and cardiac cycle, monitoring respiration, monitoring depth of anesthesia, monitoring hypo and hyper-volemia etc.

A PPG (Photoplethysmograph) probe senses the PPG signal from the flow rate of our blood and converts it into the electrical signal with the help of receiver and the source. The signal produced upto here is from 0 to 1.5 volts. The probe model number is DS-100A manufactured by Nellcor. It is a reusable sensor for spot checks or short-term monitoring. PPG probe has adult finger sensor which is used for pulse oximeters. A DB9 connector is connected to the probe which has 9 pins.

The application site of the probe includes ear, head, finger and toe. The cable length is 3ft/0.9m - 10ft/3.0m. There are various features of the PPG probe which includes:

1. Reusable adult sensor
2. Easy to use and design
3. Built in shielding protects signal from noise
4. High quality LEDs maximize tracking capabilities
5. Sensor head design optimizes signal and shields detector from ambient light.

Figure 3-4 shows the Nellcor SpO$_2$ sensor.

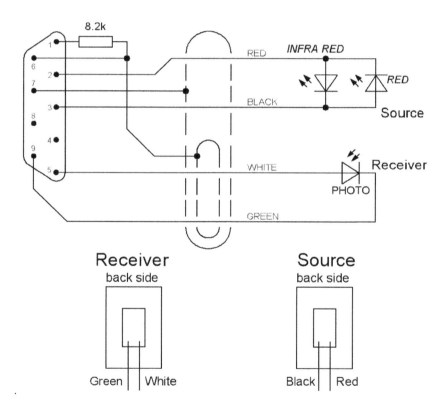

# SpO2 sensor Nellcor

Figure 3-4: Nellcor SpO$_2$ sensor

The probe has DB9 connector. The view of 9 pins of the DB9 connector for both male and female connectors is shown in figure 3-5.

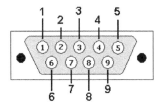

**DB9: View looking into male connector**

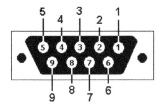

**DB9: View looking into female connector**

Figure 3-5: DB9 connector

A male connector is a connector attached to a wire, cable, or piece of hardware, having one or more exposed, unshielded electrical terminals, and constructed in such a way that it can be inserted snugly into a receptacle (female connector) to ensure a reliable physical and electrical connection. This type of connector is also known as a plug. A male connector can be recognized by the fact that, when it is disconnected or removed, the unshielded electrical prongs are plainly visible.

The most common male connector is a two- or three-prong plug attached to the end of the cord for an electrical appliance. Other common examples include the plugs for headsets, most connectors on the ends of lengths of coaxial cable , and the edge connectors on some printed circuit cards. D-shell connectors for computer serial and parallel ports can also be male.

A gender changer is a hardware device used to convert a male connector to a female connector, or vice versa.

The probe having finger-clip with the connector is shown in figure 3-6.

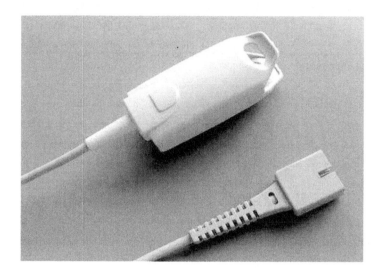

Figure 3-6: Finger-clip with the connector

There are two types of sensors namely:

1) Transmittance

2) Reflectance

Transmittance sensors must have the light source properly aligned with the photodetector whereas reflectance sensors require proper alignment of the sensor against the surface of the skin.

We are using transmittance sensor.

### 3.1.2 Amplifier stage

The amplifier stage is further divided into two stages:

1. Current to voltage (I/V) converter

2. An amplifier

**3.1.2.1 Current to voltage converter:** A current to voltage converter is an electrical device that takes an electric current as an input signal and produces a corresponding voltage as an output signal. It is also known as transimpedance amplifier. Current-to-voltage converter is a circuit that performs current to voltage transformation. In electronic circuitry operating at signal voltages, it usually changes the electric attribute carrying information from current to voltage. The converter acts as a linear circuit with transfer ratio $k = V_{OUT}/I_{IN}$, called the transimpedance, which has dimensions of [V/A] (also known as resistance). That is why the active version of the circuit is also referred to as a transresistance or transimpedance amplifier.

Typical applications of current-to-voltage converter are measuring currents by using instruments having voltage inputs, creating current-controlled voltage sources, building various passive and active voltage-to-voltage converters, etc. In some cases, the simple passive current-to-voltage converter works well; in other cases, there is a need of using active current-to-voltage converters. Ideal current-to-voltage converters have zero input resistance (impedance), so that they actually short the input source. Therefore, in this case, the input source has to have some resistance; ideally, it has to behave as a constant current source. Otherwise, the input source and the current-to-voltage converter can saturate.

I am using IC UA741 (op-amp) as a current to voltage converter. An operational amplifier is a DC-coupled high-gain electronic voltage amplifier with a differential input and, usually, a single-ended output. An op-amp produces an output voltage that is typically hundreds of thousands times larger than the voltage difference between its input terminals.

Operational amplifiers are important building blocks for a wide range of electronic circuits. They had their origins in analog computers where they were used in many linear, non-linear and frequency-dependent circuits. Their popularity in circuit design largely stems from the fact that characteristics of the final elements (such as their gain) are set by external components with little dependence on temperature changes and manufacturing variations in the op-amp itself.

Op-amps are among the most widely used electronic devices today, being used in a vast array of consumer, industrial, and scientific devices. Many standard IC op-amps cost only a few cents in moderate production volume; however some integrated or hybrid operational amplifiers with special performance specifications may cost over $100 US in small quantities. Op-amps may be packaged as components, or used as elements of more complex integrated circuits. Figure 3-7 shows IC UA741.

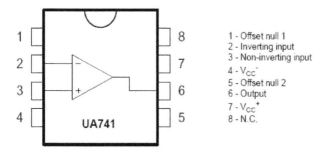

Figure 3-7: IC UA741

**3.1.2.2 An amplifier:** An amplifier amplifies the signal. An amplifier or simply amp, is a device for increasing the power of a signal. In popular use, the term usually describes an electronic amplifier, in which the input signal is usually a voltage or a current. Amplifiers may be classified according to the input (source) they are designed to amplify, the device they are intended to drive, the frequency range of the signals (Audio, IF, RF, and VHF amplifiers, for example), whether they invert the signal (inverting amplifiers and non-inverting amplifiers), or the type of device used in the amplification (FET amplifiers etc.). The quality of an amplifier can be characterized by a number of specifications like gain, bandwidth, efficiency etc.

IC UA741 is used as an amplifier with two potentiometers to adjust the dc level of the signal.

**3.1.3   ADC stage:** An analog-to-digital converter (abbreviated ADC, A/D or A to D) is a device which converts a continuous quantity to a discrete time digital representation. Typically, an ADC is an electronic device  that  converts  an  input  analog  voltage or current to  a  digital  number proportional to the magnitude of the voltage or current.

The microcontroller ADuC842 has on-chip ADC. The ADC conversion block incorporates a fast, 8-channel, 12-bit, single-supply ADC. This block provides the user with multichannel mux, track-and-hold, on-chip reference, calibration features, and ADC. All components in this block are easily configured via a 3-register SFR interface. The ADC converter consists of a conventional successive approximation converter based around a capacitor DAC. The converter accepts an analog input range of 0 V to $V_{REF}$. A high precision, 15 ppm, low drift, factory calibrated 2.5 V reference is provided on-chip. The external reference can be in the range of 1 V to $AV_{DD}$.

Single-step or continuous conversion modes can be initiated in software or alternatively by applying a convert signal to an external pin. Timer 2 can also be configured to generate a repetitive trigger for ADC conversions. The ADC may be configured to operate in a DMA mode whereby the ADC block continuously converts and captures samples to an external RAM space without any interaction from the MCU core. This automatic capture facility can extend through a 16 MByte external data memory space.

The ADuC841/ADuC842/ADuC843 is shipped with factory programmed calibration coefficients that are automatically downloaded to the ADC on power-up, ensuring optimum ADC performance. The ADC core contains internal offset and gain calibration registers that can be hardware calibrated to minimize system errors.

A voltage output from an on-chip band gap reference proportional to absolute temperature can also be routed through the front end ADC multiplexer (effectively a 9th ADC channel input), facilitating a temperature sensor implementation.

**3.1.3.1 ADC Transfer function:** The analog input range for the ADC is 0 V to $V_{REF}$. For this range, the designed code transitions occur midway between successive integer LSB values, i.e., 0.5 LSB, 1.5 LSB, 2.5 LSB . . . FS −1.5 LSB. The output coding is straight binary with 1 LSB = FS/4096 or 2.5 V/4096 = 0.61 mV when $V_{REF}$ = 2.5 V. The ideal input/output transfer characteristic for the 0 V to VREF range is shown in figure 3-8.

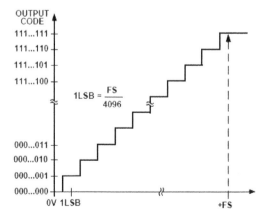

Figure 3-8: ADC Transfer Function

**3.1.3.2 Typical operation:** Once configured via the ADCCON 1–3 SFRs, the ADC converts the analog input and provides an ADC 12-bit result word in the ADCDATAH/L SFRs. The top 4 bits of the ADCDATAH SFR are written with the channel selection bits to identify the channel result. The format of the ADC 12-bit result word is shown in figure 3-9.

32

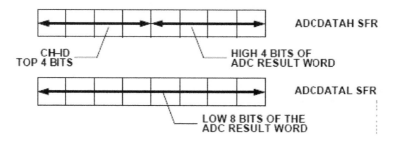

Figure 3-9: ADC result word format

The ADC incorporates a successive approximation architecture (SAR) involving a charge-sampled input stage. Each ADC conversion is divided into two distinct phases as defined by the position of the switches in Figure 3-10. During the sampling phase (with SW1 and SW2 in the track position), a charge proportional to the voltage on the analog input is developed across the input sampling capacitor. During the conversion phase (with both switches in the hold position), the capacitor DAC is adjusted via internal SAR logic until the voltage on Node A is 0, indicating that the sampled charge on the input capacitor is balanced out by the charge being output by the capacitor DAC. The final digital value contained in the SAR is then latched out as the result of the ADC conversion. Control of the SAR and timing of acquisition and sampling modes is handled automatically by built-in ADC control logic. Acquisition and conversion times are also fully configurable under user control.

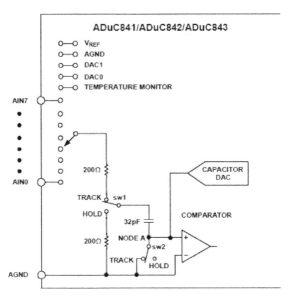

Figure 3-10: Internal ADC structure

Note that whenever a new input channel is selected, a residual charge from the 32 pF sampling capacitor places a transient on the newly selected input. The signal source must be capable of recovering from this transient before the sampling switches go into hold mode. Delays can be inserted in software (between channel selection and conversion request) to account for input stage settling, but hardware solution alleviates this burden from the software design task and ultimately results in a cleaner system implementation. One hardware solution is to choose a very fast settling op amp to drive each analog input. Such an op amp would need to fully settle from a small signal transient in less than 300 ns in order to guarantee adequate settling under all software configurations.

A better solution, recommended for use with any amplifier, is shown in figure 3-11. Though at first glance the circuit in figure 3-11 may look like a simple antialiasing filter, it actually serves no such purpose since its corner frequency is well above the Nyquist frequency, even at a 200 kHz sample rate. Though the R/C does help to reject some incoming high frequency noise, its primary function is to ensure that the transient demands of the ADC input stage are met.

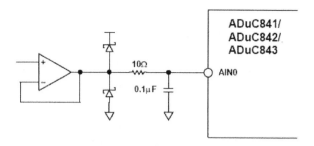

Figure 3-11: Buffering analog inputs

It does so by providing a capacitive bank from which the 32 pF sampling capacitor can draw its charge. Its voltage does not change by more than one count (1/4096) of the 12-bit transfer function when the 32 pF charge from a previous channel is dumped onto it. A larger capacitor can be used if desired, but not a larger resistor (for reasons described below). The Schottky diodes in figure 3-11 may be necessary to limit the voltage applied to the analog input pin per the Absolute Maximum Ratings. They are not necessary if the op amp is powered from the same supply as the part since in that case the op amp is unable to generate voltages above $V_{DD}$ or below ground. An op amp of some kind is necessary unless the signal source is very low impedance to begin with. DC leakage currents at the parts'analog inputs can cause measurable dc errors with external source impedances as low as 100 $\Omega$ or so. To ensure accurate ADC operation, keep the total source impedance at each analog input less than 61 $\Omega$. Table 3-1 illustrates examples of how source impedance can affect dc accuracy.

Table 3-1: Source Impedance and DC Accuracy

| Source Impedance $\Omega$ | Error from 1 µA Leakage Current | Error from 10 µA Leakage Current |
|---|---|---|
| 61 | 61 µV = 0.1 LSB | 610 µV = 1 LSB |
| 610 | 610 µV = 1 LSB | 6.1 mV = 10 LSB |

Although figure 3-11 shows the op amp operating at a gain of 1, one can, of course, configure it for any gain needed. Also, one can just as easily use an instrumentation amplifier in its place to condition differential signals. Use amplifier that is capable of delivering the signal (0 V to $V_{REF}$) with minimal saturation. Some single-supply rail-to-rail op amps that are useful for this purpose are described in table 3-2.

35

Table 3-2: Some Single-Supply Op Amps

| Op Amp Model | Characteristics |
|---|---|
| OP281/OP481 | Micropower |
| OP191/OP291/OP491 | I/O Good up to $V_{DD}$, Low Cost |
| OP196/OP296/OP496 | I/O to $V_{DD}$, Micropower, Low Cost |
| OP183/OP283 | High Gain-Bandwidth Product |
| OP162/OP262/OP462 | High GBP, Micro Package |
| AD820/AD822/AD824 | FET Input, Low Cost |
| AD823 | FET Input, High GBP |

The ADC's transfer function is 0 V to $V_{REF}$, and that any signal range lost to amplifier saturation near ground will impact dynamic range. Though the op amps in table 3-2 are capable of delivering output signals that very closely approach ground, no amplifier can deliver signals all the way to ground when powered by a single supply. Therefore, if a negative supply is available, we might consider using it to power the front end amplifiers. If we do, however, be sure to include the Schottky diodes shown in figure 3-11 (or at least the lower of the two diodes) to protect the analog input from undervoltage conditions. To summarize this section, use the circuit in figure 3-11 to drive the analog input pins of the parts.

### 3.1.3.3 Voltage reference connections

The on-chip 2.5 V band gap voltage reference can be used as the reference source for the ADC and DACs. To ensure the accuracy of the voltage reference, we must decouple the $C_{REF}$ pin to ground with a 0.47 µF capacitor as shown in figure 3-12.

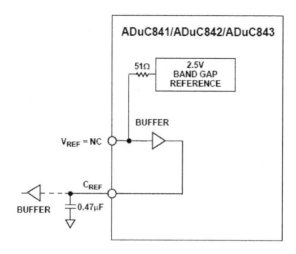

Figure 3-12: Decoupling $V_{REF}$ and $C_{REF}$

If the internal voltage reference is to be used as a reference for external circuitry, the $C_{REF}$ output should be used. However, a buffer must be used in this case to ensure that no current is drawn from the $C_{REF}$ pin itself. The voltage on the $C_{REF}$ pin is that of an internal node within the buffer block, and its voltage is critical for ADC and DAC accuracy. The parts power up with their internal voltage reference in the off state.

If an external voltage reference is preferred, it should be connected to the $C_{REF}$ pin as shown in Figure 3-13. Bit 6 of the ADCCON1 SFR must be set to 1 to switch in the external reference voltage.

To ensure accurate ADC operation, the voltage applied to $C_{REF}$ must be between 1 V and $AV_{DD}$. In situations where analog input signals are proportional to the power supply (such as in some strain gage applications), it may be desirable to connect the CREF pin directly to $AV_{DD}$. Operation of the ADC or DACs with a reference voltage below 1 V, however, may incur loss of accuracy, eventually resulting in missing codes or nonmonotonicity. For that reason, do not use a reference voltage lower than 1 V.

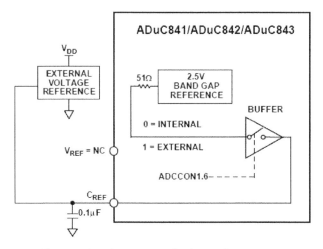

Figure 3-13: Using an external voltage reference

To summarize, the ADCCON1 and ADCCON2 are two SFRs that need to be configured while writing the program. The ADC converter consists of a conventional successive approximation converter. A 2.5V is provided on-chip. The converter accepts an analog input range of 0V to $V_{REF}$. The external reference can be in the range of 1V to $AV_{DD}$.

### 3.1.3.4 ADC Control Special Function Registers (SFRs):

There are three ADC Control Special Function Registers namely ADCCON1, ADCCON2 and ADCCON3. Each one of these is described below.

### (i)     ADCCON1—(ADC Control SFR 1)

The ADCCON1 register controls conversion and acquisition times, hardware conversion modes, and power-down modes as detailed below:

SFR Address: EFH

SFR Power-On Default: 40H

Bit Addressable: No

The ADCCON1 SFR bit designations are given in table 3-3.

## Table 3-3: ADCCON1 SFR bit designations

| Bit No. | Name | Description |
|---------|------|-------------|
| 7 | MD1 | The mode bit selects the active operating mode of the ADC.<br>Set by the user to power up the ADC.<br>Cleared by the user to power down the ADC. |
| 6 | EXT_REF | Set by the user to select an external reference.<br>Cleared by the user to use the internal reference. |
| 5<br>4 | CK1<br>CK0 | The ADC clock divide bits (CK1, CK0) select the divide ratio for the PLL master clock (ADuC842/ADuC843) or the external crystal (ADuC841) used to generate the ADC clock. To ensure correct ADC operation, the divider ratio must be chosen to reduce the ADC clock to 8.38 MHz or lower. A typical ADC conversion requires 16 ADC clocks plus the selected acquisition time.<br>The divider ratio is selected as follows:<br><br>CK1   CK0   MCLK Divider<br>0     0     32<br>0     1     4 (Do not use with a CD setting of 0)<br>1     0     8<br>1     1     2 |
| 3<br>2 | AQ1<br>AQ0 | The ADC acquisition select bits (AQ1, AQ0) select the time provided for the input track-and-hold amplifier to acquire the input signal. An acquisition of three or more ADC clocks is recommended; clocks are as follows:<br><br>AQ1   AQ0   No. ADC Clks<br>0     0     1<br>0     1     2<br>1     0     3<br>1     1     4 |
| 1 | T2C | The Timer 2 conversion bit (T2C) is set by the user to enable the Timer 2 overflow bit to be used as the ADC conversion start trigger input. |
| 0 | EXC | The external trigger enable bit (EXC) is set by the user to allow the external Pin P3.5 ($\overline{\text{CONVST}}$) to be used as the active low convert start input. This input should be an active low pulse (minimum pulse width >100 ns) at the required sample rate. |

## (ii)    ADCCON2—(ADC Control SFR 2)

The ADCCON2 register controls ADC channel selection and conversion modes as detailed below:

SFR Address: D8H

SFR Power-On Default: 00H

Bit Addressable: Yes

The ADCCON2 SFR bit designations are given in table 3-4.

Table 3-4: ADCCON2 SFR bit designations

| Bit No. | Name | Description |
|---|---|---|
| 7 | ADCI | ADC Interrupt Bit. |
|  |  | Set by hardware at the end of a single ADC conversion cycle or at the end of a DMA block conversion. |
|  |  | Cleared by hardware when the PC vectors to the ADC interrupt service routine. Otherwise, the ADCI bit is cleared by user code. |
| 6 | DMA | DMA Mode Enable Bit. |
|  |  | Set by the user to enable a preconfigured ADC DMA mode operation. A more detailed description of this mode is given in the ADC DMA Mode section. The DMA bit is automatically set to 0 at the end of a DMA cycle. Setting this bit causes the ALE output to cease; it will start again when DMA is started and will operate correctly after DMA is complete. |
| 5 | CCONV | Continuous Conversion Bit. |
|  |  | Set by the user to initiate the ADC into a continuous mode of conversion. In this mode, the ADC starts converting based on the timing and channel configuration already set up in the ADCCON SFRs; the ADC automatically starts another conversion once a previous conversion has completed. |
| 4 | SCONV | Single Conversion Bit. |
|  |  | Set to initiate a single conversion cycle. The SCONV bit is automatically reset to 0 on completion of the single conversion cycle. |
| 3 | CS3 | Channel Selection Bits. |
| 2 | CS2 | Allow the user to program the ADC channel selection under software control. When a conversion is initiated, the |
| 1 | CS1 | converted channel is the one pointed to by these channel selection bits. In DMA mode, the channel selection is |
| 0 | CS0 | derived from the channel ID written to the external memory. |

| CS3 | CS2 | CS1 | CS0 | CH# | |
|---|---|---|---|---|---|
| 0 | 0 | 0 | 0 | 0 | |
| 0 | 0 | 0 | 1 | 1 | |
| 0 | 0 | 1 | 0 | 2 | |
| 0 | 0 | 1 | 1 | 3 | |
| 0 | 1 | 0 | 0 | 4 | |
| 0 | 1 | 0 | 1 | 5 | |
| 0 | 1 | 1 | 0 | 6 | |
| 0 | 1 | 1 | 1 | 7 | |
| 1 | 0 | 0 | 0 | Temp Monitor | Requires minimum of 1 μs to acquire. |
| 1 | 0 | 0 | 1 | DAC0 | Only use with internal DAC output buffer on. |
| 1 | 0 | 1 | 0 | DAC1 | Only use with internal DAC output buffer on. |
| 1 | 0 | 1 | 1 | AGND | |
| 1 | 1 | 0 | 0 | $V_{REF}$ | |
| 1 | 1 | 1 | 1 | DMA STOP | Place in XRAM location to finish DMA sequence; refer to the ADC DMA Mode section. |

All other combinations reserved.

## (iii) ADCCON3—(ADC Control SFR 3)

The ADCCON3 register controls the operation of various calibration modes and also indicates the ADC busy status.

SFR Address: F5H

SFR Power-On Default: 00H

Bit Addressable: No

The ADCCON3 SFR bit designations are given in table 3-5.

40

## Table 3-5: ADCCON3 SFR bit designations

| Bit No. | Name | Description |
|---|---|---|
| 7 | BUSY | ADC Busy Status Bit.<br>A read-only status bit that is set during a valid ADC conversion or during a calibration cycle.<br>Busy is automatically cleared by the core at the end of conversion or calibration. |
| 6 | RSVD | Reserved. This bit should always be written as 0. |
| 5 | AVGS1 | Number of Average Selection Bits. |
| 4 | AVGS0 | This bit selects the number of ADC readings that are averaged during a calibration cycle.<br><br>AVGS1    AVGS0    Number of Averages<br>0    0    15<br>0    1    1<br>1    0    31<br>1    1    63 |
| 3 | RSVD | Reserved. This bit should always be written as 0. |
| 2 | RSVD | This bit should always be written as 1 by the user when performing calibration. |
| 1 | TYPICAL | Calibration Type Select Bit.<br>This bit selects between offset (zero-scale) and gain (full-scale) calibration.<br>Set to 0 for offset calibration.<br>Set to 1 for gain calibration. |
| 0 | SCAL | Start Calibration Cycle Bit.<br>When set, this bit starts the selected calibration cycle.<br>It is automatically cleared when the calibration cycle is completed. |

### 3.1.3.5 Configuring the ADC

The parts' successive approximation ADC is driven by a divided down version of the master clock. To ensure adequate ADC operation, this ADC clock must be between 400 kHz and 8.38 MHz. Frequencies within this range can be achieved easily with master clock frequencies from 400 kHz to well above 16 MHz, with the four ADC clock divide ratios to choose from. For example, set the ADC clock divide ratio to 8 (i.e., ADCCLK = 16.777216 MHz/8 = 2 MHz) by setting the appropriate bits in ADCCON1 (ADCCON1.5 = 1, ADCCON1.4 = 0). The total ADC conversion time is 15 ADC clocks, plus 1 ADC clock for synchronization, plus the selected acquisition time (1, 2, 3, or 4 ADC clocks). For the preceding example, with a 3-clock acquisition time, total conversion time is 19 ADC clocks (or 9.05 μs for a 2 MHz ADC clock).

In continuous conversion mode, a new conversion begins each time the previous one finishes. The sample rate is then simply the inverse of the total conversion time described previously. In the preceding example, the continuous conversion mode sample rate is 110.3 kHz.

If using the temperature sensor as the ADC input, the ADC should be configured to use an ADCCLK of MCLK/32 and four acquisition clocks.

Increasing the conversion time on the temperature monitor channel improves the accuracy of the reading. To further improve the accuracy, an external reference with low temperature drift should also be used.

### 3.1.3.6 ADC Offset and Gain Calibration Coefficients

The ADuC841/ADuC842/ADuC843 has two ADC calibration coefficients, one for offset calibration and one for gain calibration. Both the offset and gain calibration coefficients are 14-bit words, and are each stored in two registers located in the special function register (SFR) area. The offset calibration coefficient is divided into ADCOFSH (six bits) and ADCOFSL (8 bits), and the gain calibration coefficient is divided into ADCGAINH (6 bits) and ADCGAINL (8 bits).

The offset calibration coefficient compensates for dc offset errors in both the ADC and the input signal. Increasing the offset coefficient compensates for positive offset, and effectively pushes the ADC transfer function down. Decreasing the offset coefficient compensates for negative offset, and effectively pushes the ADC transfer function up. The maximum offset that can be compensated is typically $\pm5\%$ of $V_{REF}$, which equates to typically $\pm125$ mV with a 2.5 V reference.

Similarly, the gain calibration coefficient compensates for dc gain errors in both the ADC and the input signal. Increasing the gain coefficient compensates for a smaller analog input signal range and scales the ADC transfer function up, effectively increasing the slope of the transfer function. Decreasing the gain coefficient compensates for a larger analog input signal range and scales the ADC transfer function down, effectively decreasing the slope of the transfer function. The maximum analog input signal range for which the gain coefficient can compensate is $1.025 \times V_{REF}$, and the minimum input range is $0.975 \times V_{REF}$, which equates to typically $\pm2.5\%$ of the reference voltage.

### 3.1.3.7 Calibrating the ADC

Two hardware calibration modes are provided, which can be easily initiated by user software. The ADCCON3 SFR is used to calibrate the ADC. Bit 1 (typical) and CS3 to CS0 (ADCCON2) set up the calibration modes.

Device calibration can be initiated to compensate for significant changes in operating condition frequency, analog input range, reference voltage, and supply voltages. In this calibration mode, offset calibration uses internal AGND selected via ADCCON2 register Bits CS3 to CS0 (1011), and gain calibration uses internal $V_{REF}$ selected by Bits CS3 to CS0 (1100). Offset calibration should be executed first, followed by gain calibration. System calibration can be initiated to compensate for both internal and external system errors. To perform system calibration by using an external reference, tie the system ground and reference to any two of the six selectable inputs. Enable external reference mode (ADCCON1.6). Select the channel connected to AGND via Bits CS3 to

CS0 and perform system offset calibration. Select the channel connected to $V_{REF}$ via Bits CS3 to CS0 and perform system gain calibration.

### 3.1.3.8 Initiating the Calibration in Code

When calibrating the ADC using ADCCON1, the ADC must be set up into the configuration in which it will be used. The ADCCON3 register can then be used to set up the device and to calibrate the ADC offset and gain.

```
MOV ADCCON1, #08CH        ; ADC on; ADCCLK set
                          ; to divide by 32,4
                          ; acquisition clock
To calibrate device offset:
MOV ADCCON2, #0BH         ; select internal AGND
MOV ADCCON3, #25H         ; select offset calibration,
                          ; 31 averages per bit,
                          ; offset calibration
To calibrate device gain:
MOV ADCCON2, #0CH         ; select internal VREF
MOV ADCCON3, #27H         ; select offset calibration,
                          ; 31 averages per bit,
                          ; offset calibration
To calibrate system offset, connect system AGND to an ADC channel input (0).
MOV ADCCON2, #00H         ; select external AGND
MOV ADCCON3, #25H         ; select offset calibration,
                          ; 31 averages per bit
To calibrate system gain, connect system VREF to an ADC channel input (1).
MOV ADCCON2, #01H         ; select external VREF
MOV ADCCON3, #27H         ; select offset calibration,
                          ; 31 averages per bit,
                          ; offset calibration
```

The calibration cycle time $T_{CAL}$ is calculated by the following equation:

$$T_{CAL} = 14 \times ADCCLK \times NUMAV \times (16 + T_{ACQ})$$

For an ADCCLK/FCORE divide ratio of 32, $T_{ACQ}$ = 4 ADCCLK, and NUMAV = 15, the calibration cycle time is

$$T_{CAL} = 14 \times (1/524288) \times 15 \times (16+4)$$
$$T_{CAL} = 8\ ms$$

In a calibration cycle, the ADC busy flag (Bit 7), instead of framing an individual ADC conversion as in normal mode, goes high at the start of calibration and returns to zero only at the end of the calibration cycle. It can therefore be monitored in code to indicate when the calibration cycle is completed. The following code can be used to monitor the BUSY signal during a calibration cycle:

```
WAIT:
MOV A, ADCCON3          ; move ADCCON3 to A
JB ACC.7, WAIT          ; if Bit 7 is set jump to WAIT else continue
```

**3.1.4  Microcontroller:** The microcontroller ADuC842 has the following features:

1. Pin compatable ugrade of ADuC812/ADuC831/ADuC832.

2. Increased performance (single-cycle 20 MIPS 8052 core and high speed 420 kSPS 12-bit ADC).

3. Increased memory (up to 62 kBytes on-chip Flash/EE program memory and 4 kBytes on-chip Flash/EE data memory).

4. In-circuit reprogrammable (Flash/EE, 100 year retention, 100 kCycle endurance and 2304 bytes on-chip data RAM).

5. Smaller package i.e. 8 mm × 8 mm chip scale package and 52-lead PQFP—pin compatable upgrade.

6. Analog I/O (8-channel, 420 kSPS high accuracy, 12-bit ADC; on-chip, 15 ppm/°C voltage reference; DMA controller, high speed ADC-to-RAM capture; two 12-bit voltage output DACs; dual output PWM Σ-Δ DACs and on-chip temperature monitor function).

7. 8052 based core and 8051 compatible instruction set (20 MHz max).

8. High performance single-cycle core (32 kHz external crystal, on-chip programmable PLL 12 interrupt sources, 2 priority levels; dual data pointers, extended 11-bit stack pointer).

9. On-chip peripherals (Time Interval Counter (TIC), UART, I2CR, and SPIR Serial I/O watchdog timer (WDT) and power supply monitor (PSM)).

10. Power (Normal: 4.5 mA @ 3 V (core CLK = 2.098 MHz) and power-down: 10 μA @ 3 V).

11. Development tools like low cost, comprehensive development system, incorporating nonintrusive single-pin emulation, IDE based assembly and C source debugging.

The functional block diagram of ADuC842 microcontroller is shown in figure 3-14 and its pin diagram is shown in figure 3-15.

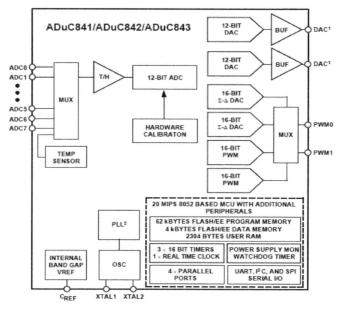

Figure 3-14: Functional block diagram of ADuC842 microcontroller

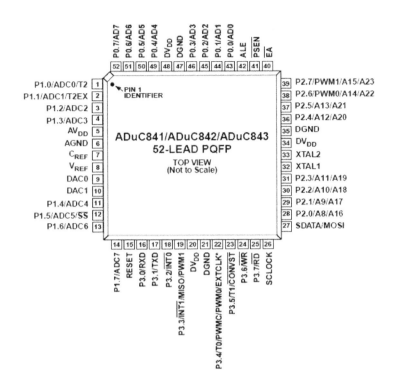

Figure 3-15: Pin diagram of ADuC842 microcontroller

The detailed pin description of ADuC842 microcontroller is given in table 3-6.

## Table 3-6: Pin description of ADuC842 microcontroller

| Mnemonic | Type | Function |
|---|---|---|
| DV$_{DD}$ | P | Digital Positive Supply Voltage. 3 V or 5 V nominal. |
| AV$_{DD}$ | P | Analog Positive Supply Voltage. 3 V or 5 V nominal. |
| C$_{REF}$ | I/O | Decoupling Input for On-Chip Reference. Connect a 0.47 µF capacitor between this pin and AGND. |
| V$_{REF}$ | NC | Not connected. This was reference out on the ADuC812; the C$_{REF}$ pin should be used instead. |
| AGND | G | Analog Ground. Ground reference point for the analog circuitry. |
| P1.0–P1.7 | I | Port 1 is an 8-bit input port only. Unlike other ports, Port 1 defaults to analog input mode. To configure any of these port pins as a digital input, write a 0 to the port bit. |
| ADC0–ADC7 | I | Analog Inputs. Eight single-ended analog inputs. Channel selection is via ADCCON2 SFR. |
| T2 | I | Timer 2 Digital Input. Input to Timer/Counter 2. When enabled, Counter 2 is incremented in response to a 1-to-0 transition of the T2 input. |
| T2EX | I | Digital Input. Capture/reload trigger for Counter 2; also functions as an up/down control input for Counter 2. |
| $\overline{SS}$ | I | Slave Select Input for the SPI Interface. |
| SDATA | I/O | User Selectable, I²C Compatible, or SPI Data Input/Output Pin. |
| SCLOCK | I/O | Serial Clock Pin for I²C Compatible or for SPI Serial Interface Clock. |
| MOSI | I/O | SPI Master Output/Slave Input Data I/O Pin for SPI Interface. |
| MISO | I/O | SPI Master Input/Slave Output Data I/O Pin for SPI Serial Interface. |
| DAC0 | O | Voltage Output from DAC0. This pin is a no connect on the ADuC843. |
| DAC1 | O | Voltage Output from DAC1. This pin is a no connect on the ADuC843. |
| RESET | I | Digital Input. A high level on this pin for 24 master clock cycles while the oscillator is running resets the device. |

47

| Mnemonic | Type | Function |
|---|---|---|
| P3.0–P3.7 | I/O | Port 3 is a bidirectional port with internal pull-up resistors. Port 3 pins that have 1s written to them are pulled high by the internal pull-up resistors, and in that state can be used as inputs. As inputs, Port 3 pins being pulled externally low source current because of the internal pull-up resistors. Port 3 pins also contain various secondary functions, which are described below. |
| PWMC | I | PWM Clock Input. |
| PWM0 | O | PWM0 Voltage Output. PWM outputs can be configured to use Ports 2.6 and 2.7 or Ports 3.4 and 3.3. |
| PWM1 | O | PWM1 Voltage Output. See the CFG841/CFG842 register for further information. |
| RxD | I/O | Receiver Data Input (Asynchronous) or Data Input/Output (Synchronous) of the Serial (UART) Port. |
| TxD | O | Transmitter Data Output (Asynchronous) or Clock Output (Synchronous) of the Serial (UART) Port. |
| $\overline{INT0}$ | I | Interrupt 0. Programmable edge or level triggered interrupt input; can be programmed to one of two priority levels. This pin can also be used as a gate control input to Timer 0. |
| $\overline{INT1}$ | I | Interrupt 1. Programmable edge or level triggered interrupt input; can be programmed to one of two priority levels. This pin can also be used as a gate control input to Timer 1. |
| T0 | I | Timer/Counter 0 Input. |
| T1 | I | Timer/Counter 1 Input. |
| $\overline{CONVST}$ | I | Active Low Convert Start Logic Input for the ADC Block when the External Convert Start Function is Enabled. A low-to-high transition on this input puts the track-and-hold into hold mode and starts the conversion. |
| EXTCLK | I | Input for External Clock Signal. Has to be enabled via the CFG842 register. |
| $\overline{WR}$ | O | Write Control Signal, Logic Output. Latches the data byte from Port 0 into the external data memory. |
| $\overline{RD}$ | O | Read Control Signal, Logic Output. Enables the external data memory to Port 0. |
| XTAL2 | O | Output of the Inverting Oscillator Amplifier. |
| XTAL1 | I | Input to the Inverting Oscillator Amplifier. |
| DGND | G | Digital Ground. Ground reference point for the digital circuitry. |
| P2.0–P2.7 (A8–A15) (A16–A23) | I/O | Port 2 is a bidirectional port with internal pull-up resistors. Port 2 pins that have 1s written to them are pulled high by the internal pull-up resistors, and in that state can be used as inputs. As inputs, Port 2 pins being pulled externally low source current because of the internal pull-up resistors. Port 2 emits the middle and high-order address bytes during accesses to the external 24-bit external data memory space. |
| $\overline{PSEN}$ | O | Program Store Enable, Logic Output. This pin remains low during internal program execution. $\overline{PSEN}$ is used to enable serial download mode when pulled low through a resistor on power-up or reset. On reset this pin will momentarily become an input and the status of the pin is sampled. If there is no pulldown resistor in place the pin will go momentarily high and then user code will execute. If a pull-down resistor is in place, the embedded serial download/debug kernel will execute. |
| ALE | O | Address Latch Enable, Logic Output. This output is used to latch the low byte and page byte for 24-bit address space accesses of the address into external data memory. |
| $\overline{EA}$ | I | External Access Enable, Logic Input. When held high, this input enables the device to fetch code from internal program memory locations. The parts do not support external code memory. This pin should not be left floating. |
| P0.7–P0.0 (A0-A7) | I/O | Port 0 is an 8-bit open-drain bidirectional I/O port. Port 0 pins that have 1s written to them float, and in that state can be used as high impedance inputs. Port 0 is also the multiplexed low-order address and data bus during accesses to external data memory. In this application, it uses strong internal pull-ups when emitting 1s. |

Types: P = Power, G = Ground, I = Input, O = Output, NC = No Connect

The ADuC842 are complete smart transducer front ends, that integrates a high performance self calibrating multichannel ADC, a dual DAC, and an optimized single-cycle 20 MHz 8-bit MCU (8051 instruction set compatible) on a single chip.

48

The ADuC841 and ADuC842 are identical with the exception of the clock oscillator circuit; the ADuC841 is clocked directly from an external crystal up to 20 MHz whereas the ADuC842 uses a 32 kHz crystal with an on-chip PLL generating a programmable core clock up to 16.78 MHz. The ADuC843 is identical to the ADuC842 except that the ADuC843 has no analog DAC outputs.

The microcontroller is an optimized 8052 core offering up to 20 MIPS peak performance. Three different memory options are available offering up to 62 kBytes of nonvolatile Flash/EE program memory. Four kBytes of nonvolatile Flash/EE data memory, 256 bytes RAM, and 2 kBytes of extended RAM are also integrated on-chip.

The parts also incorporate additional analog functionality with two 12-bit DACs, power supply monitor, and a band gap reference. On-chip digital peripherals include two 16-bit $\Sigma$-$\Delta$ DACs, a dual output 16-bit PWM, a watchdog timer, a time interval counter, three timers/counters, and three serial I/O ports (SPI, I2C, and UART).

On the ADuC812 and the ADuC832, the I2C and SPI interfaces share some of the same pins. For backwards compatibility, this is also the case for the ADuC841/ADuC842/ADuC843.

However, there is also the option to allow SPI operate separately on P3.3, P3.4, and P3.5, while I2C uses the standard pins. The I2C interface has also been enhanced to offer repeated start, general call, and quad addressing.

On-chip factory firmware supports in-circuit serial download and debug modes (via UART) as well as single-pin emulation mode via the EA pin. A functional block diagram of the parts is shown in figure 3-14.

Figure 3-16 shows the ADuC842 microcontroller card.

Figure 3-16: ADuC842 microcontroller card

The detailed description of ADuC842 microcontroller is given below:

i. ALE: The output on the ALE pin on a standard 8052 part is a clock at 1/6th of the core operating frequency. On the ADuC841/ ADuC842/ADuC843, the ALE pin operates as follows. For a single machine cycle instructions, ALE is high for the first half of the machine cycle and low for the second half. The ALE output is at the core operating frequency. For a two or more machine cycle instruction, ALE is high for the first half of the first machine cycle and low for the rest of the machine cycles.

ii. External memory access: There is no support for external program memory access on the parts. When accessing external RAM, the EWAIT register may need to be programmed to give extra machine cycles to MOVX commands. This is to account for differing external RAM access speeds.

iii. EWAIT SFR: SFR Address: 9FH

Power-On Default: 00H

Bit Addressable: No

This special function register (SFR) is programmed with the number of wait states for a MOVX instruction. This value can range from 0H to 7H.

50

iv. Memory organization: The ADuC841/ADuC842/ADuC843 each contain four different memory blocks:

   a) Up to 62 kBytes of on-chip Flash/EE program memory
   b) 4 kBytes of on-chip Flash/EE data memory
   c) 256 bytes of general-purpose RAM
   d) Internal XRAM
   e) External XRAM

a) Flash/EE program memory: The parts provide up to 62 kBytes of Flash/EE program memory to run user code. The user can run code from this internal memory only. Unlike the ADuC812, where code execution can overflow from the internal code space to external code space once the PC becomes greater than 1FFFH, the parts do not support the roll-over from F7FFH in internal code space to F800H in external code space. Instead, the 2048 bytes between F800H and FFFFH appear as NOP instructions to user code. This internal code space can be downloaded via the UART serial port while the device is in-circuit. 56 kBytes of the program memory can be reprogrammed during run time; thus the code space can be upgraded in the field by using a user defined protocol, or it can be used as a data memory. For the 32 kBytes memory model, the top 8 kBytes function as the ULOAD space.

b) Flash/EE data memory: 4 kBytes of Flash/EE data memory are available to the user and can be accessed indirectly via a group of control registers mapped into the special function register (SFR) area.

c) General purpose RAM: The general-purpose RAM is divided into two separate memories: the upper and the lower 128 bytes of RAM. The lower 128 bytes of RAM can be accessed through direct or indirect addressing. The upper 128 bytes of RAM can be accessed only through indirect addressing because it shares the same address space as the SFR space, which can be accessed only through direct addressing.

The lower 128 bytes of internal data memory are mapped as shown in figure 3-17. The lowest 32 bytes are grouped into four banks of eight registers addressed as R0 to R7. The next 16 bytes (128 bits), locations 20H to 2FH above the register banks, form a block of directly addressable bit locations at Bit Addresses 00H to 7FH. The stack can be located anywhere in the internal memory address space, and the stack depth can be expanded up to 2048 bytes.

51

Reset initializes the stack pointer to location 07H and increments it once before loading the stack to start from location 08H, which is also the first register (R0) of register bank 1. Thus, if the user needs to use more than one register bank, the stack pointer should be initialized to an area of RAM not used for data storage.

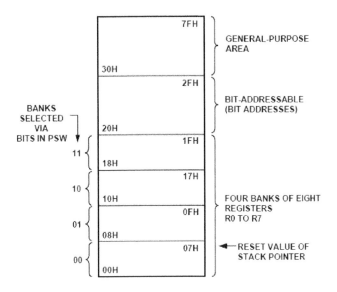

Figure 3-17: Lower 128 bytes of internal data memory

The parts contain 2048 bytes of internal XRAM, 1792 bytes of which can be configured to an extended 11-bit stack pointer.

By default, the stack operates exactly like an 8052 in that it rolls over from FFH to 00H in the general-purpose RAM. On the parts, however, it is possible (by setting CFG841.7 or CFG842.7) to enable the 11-bit extended stack pointer. In this case, the stack rolls over from FFH in RAM to 0100H in XRAM.

The 11-bit stack pointer is visible in the SP and SPH SFRs. The SP SFR is located at 81H as with a standard 8052. The SPH SFR is located at B7H. The 3 LSBs of this SFR contain the 3 extra bits necessary to extend the 8-bit stack pointer into an 11-bit stack pointer. This is shown in figure 3-18.

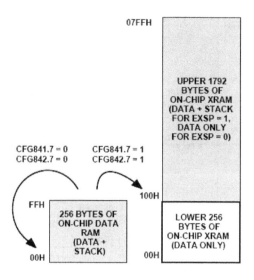

Figure 3-18: Extended stack pointer operation

d) External data memory (External XRAM): Just like a standard 8051 compatible core, the ADuC841/ADuC842/ADuC843 can access external data memory by using a MOVX instruction. The MOVX instruction automatically outputs the various control strobes required to access the data memory.

The parts, however, can access up to 16 MBytes of external data memory. This is an enhancement of the 64 kBytes of external data memory space available on a standard 8051 compatible core.

e) Internal XRAM: The parts contain 2 kBytes of on-chip data memory. This memory, although on-chip, is also accessed via the MOVX instruction. The 2 kBytes of internal XRAM are mapped into the bottom 2 kBytes of the external address space if the CFG841/CFG842 bit is set. Otherwise, access to the external data memory occurs just like a standard 8051. When using the internal XRAM, Ports 0 and 2 are free to be used as general purpose I/O. This is shown in figure 3-19.

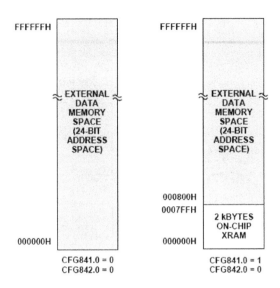

Figure 3-19: Internal and external XRAM

(v) Special Function Registers (SFRs): The SFR space is mapped into the upper 128 bytes of internal data memory space and is accessed by direct addressing only. It provides an interface between the CPU and all on-chip peripherals. A block diagram showing the programming model of the parts via the SFR area is shown in figure 3-20.

All registers, except the program counter (PC) and the four general-purpose register banks, reside in the SFR area. The SFR registers include control, configuration, and data registers, which provide an interface between the CPU and all on-chip peripherals.

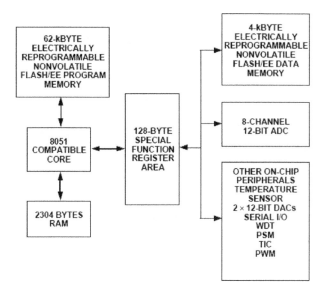

Figure 3-20: Programming model

a) Accumulator SFR (ACC): ACC is the accumulator register and is used for math operations including addition, subtraction, integer multiplication and division, and Boolean bit manipulations. The mnemonics for accumulator-specific instructions refer to the accumulator as A.

b) B SFR (B): The B register is used with the ACC for multiplication and division operations. For other instructions, it can be treated as a general-purpose scratchpad register.

c) Stack Pointer (SP and SPH): The SP SFR is the stack pointer and is used to hold an internal RAM address that is called the top of the stack. The SP register is incremented before data is stored during PUSH and CALL executions. While the stack may reside anywhere in on-chip RAM, the SP register is initialized to 07H after a reset, which causes the stack to begin at location 08H.
The parts offer an extended 11-bit stack pointer. The 3 extra bits used to make up the 11-bit stack pointer are the 3 LSBs of the SPH byte located at B7H.

d) Data Pointer (DPTR): The data pointer is made up of three 8-bit registers named DPP (page byte), DPH (high byte), and DPL (low byte). These are used to provide memory addresses for internal and external code access and for external data access. They may be manipulated

55

as a 16-bit register (DPTR = DPH, DPL), although INC DPTR instructions automatically carry over to DPP, or as three independent 8-bit registers (DPP, DPH, DPL). The parts support dual data pointers.

e) Program Status Word (PSW): The PSW SFR contains several bits reflecting the current status of the CPU.

SFR Address: D0H

Power-On Default: 00H

Bit Addressable: Yes

Table 3-7 shows bit designations of PSW SFR.

Table 3-7: PSW SFR bit designations

| Bit | Name | Description | | |
|-----|------|-------------|---|---|
| 7 | CY | Carry Flag. | | |
| 6 | AC | Auxiliary Carry Flag. | | |
| 5 | F0 | General-Purpose Flag. | | |
| 4 | RS1 | Register Bank Select Bits. | | |
| 3 | RS0 | RS1 | RS0 | Selected Bank |
| | | 0 | 0 | 0 |
| | | 0 | 1 | 1 |
| | | 1 | 0 | 2 |
| | | 1 | 1 | 3 |
| 2 | OV | Overflow Flag. | | |
| 1 | F1 | General-Purpose Flag. | | |
| 0 | P | Parity Bit. | | |

f) Power Control SFR (PCON): The PCON SFR contains bits for power-saving options and general-purpose status flags.

SFR Address: 87H

Power-On Default: 00H

Bit Addressable: No

Table 3-8 shows bit designations of PCON SFR.

Table 3-8: PCON SFR bit designations

| Bit No. | Name | Description |
|---------|------|-------------|
| 7 | SMOD | Double UART Baud Rate. |
| 6 | SERIPD | I²C/SPI Power-Down Interrupt Enable. |
| 5 | INT0PD | $\overline{INT0}$ Power-Down Interrupt Enable. |
| 4 | ALEOFF | Disable ALE Output. |
| 3 | GF1 | General-Purpose Flag Bit. |
| 2 | GF0 | General-Purpose Flag Bit. |
| 1 | PD | Power-Down Mode Enable. |
| 0 | IDL | Idle Mode Enable. |

(vi) Time Interval Counter (TIC)

A TIC is provided on-chip for counting longer intervals than the standard 8051 compatible timers are capable of. The TIC is capable of timeout intervals ranging from 1/128 second to 255 hours. Furthermore, this counter is clocked by the external 32.768 kHz crystal rather than by the core clock, and it has the ability to remain active in power-down mode and time long power-down intervals. This has obvious applications for remote battery-powered sensors where regular widely spaced readings are required.

Six SFRs are associated with the time interval counter, TIMECON being its control register. Depending on the configuration of the IT0 and IT1 bits in TIMECON, the selected time counter register overflow clocks the interval counter. When this counter is equal to the time interval value loaded in the INTVAL SFR, the TII bit (TIMECON.2) is set and generates an interrupt if enabled. If the part is in power-down mode, again with TIC interrupt enabled, the TII bit wakes up the device and resumes code execution by vectoring directly to the TIC interrupt service vector address at 0053H. The time based SFRs can be written initially with the current time; the TIC can then be controlled and accessed by user software. In effect, this facilitates the implementation of a real-time clock. A block diagram of the TIC is shown in figure 3-21.

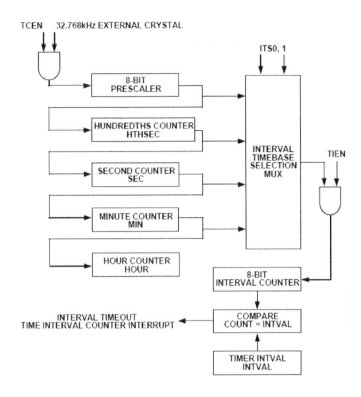

Figure 3-21: Block diagram of TIC

The TIC is clocked directly from a 32 kHz external crystal on the ADuC842/ADuC843 and by the internal 32 kHz ±10% R/C oscillator on the ADuC841. Due to this, instructions that access the TIC registers will also be clocked at this speed. The user should ensure that there is sufficient time between instructions to these registers to allow them to execute correctly.

**TIMECON:** TIC Control Register
SFR Address: A1H
Power-On Default: 00H
Bit Addressable: No

TIMECON SFR bit designations is given in table 3-9.

Table 3-9: TIMECON SFR bit designations

| Bit No. | Name | Description |
|---------|------|-------------|
| 7 | ---- | Reserved. |
| 6 | TFH | Twenty-Four Hour Select Bit. |
| | | Set by the user to enable the hour counter to count from 0 to 23. |
| | | Cleared by the user to enable the hour counter to count from 0 to 255. |
| 5 | ITS1 | Interval Timebase Selection Bits. |
| 4 | ITS0 | Written by user to determine the interval counter update rate. |

| ITS1 | ITS0 | Interval Timebase |
|------|------|-------------------|
| 0 | 0 | 1/128 Second |
| 0 | 1 | Seconds |
| 1 | 0 | Minutes |
| 1 | 1 | Hours |

| Bit No. | Name | Description |
|---------|------|-------------|
| 3 | STI | Single Time Interval Bit. |
| | | Set by the user to generate a single interval timeout. If set, a timeout clears the TIEN bit. |
| | | Cleared by the user to allow the interval counter to be automatically reloaded and start counting again at each interval timeout. |
| 2 | TII | TIC Interrupt Bit. |
| | | Set when the 8-bit interval counter matches the value in the INTVAL SFR. |
| | | Cleared by user software. |
| 1 | TIEN | Time Interval Enable Bit. |
| | | Set by the user to enable the 8-bit time interval counter. |
| | | Cleared by the user to disable the interval counter. |
| 0 | TCEN | Time Clock Enable Bit. |
| | | Set by the user to enable the time clock to the time interval counters. |
| | | Cleared by the user to disable the clock to the time interval counters and reset the time interval SFRs to the last value written to them by the user. The time registers (HTHSEC, SEC, MIN, and HOUR) can be written while TCEN is low. |

The various registers of TIC are:

a) INTVAL (User Time Interval Select Register): User code writes the required time interval to this register. When the 8-bit interval counter is equal to the time interval value loaded in the INTVAL SFR, the TII bit (TIMECON.2) is set and generates an interrupt if enabled.

   SFR Address: A6H

   Power-On Default: 00H

   Bit Addressable: No

   Valid Value: 0 to 255 decimal

b) HTHSEC (Hundredths Seconds Time Register): This register is incremented in 1/128 second intervals once TCEN in TIMECON is active. The HTHSEC SFR counts from 0 to 127 before rolling over to increment the SEC time register.

   SFR Address: A2H

   Power-On Default: 00H

   Bit Addressable: No

Valid Value: 0 to 127 decimal

c) SEC (Seconds Time Register): This register is incremented in 1-second intervals once TCEN in TIMECON is active. The SEC SFR counts from 0 to 59 before rolling over to increment the MIN time register.
SFR Address: A3H
Power-On Default: 00H
Bit Addressable: No
Valid Value: 0 to 59 decimal

d) MIN (Minutes Time Register): This register is incremented in 1-minute intervals once TCEN in TIMECON is active. The MIN SFR counts from 0 to 59 before rolling over to increment the HOUR time register.
SFR Address: A4H
Power-On Default: 00H
Bit Addressable: No
Valid Value: 0 to 59 decimal

e) HOUR (Hours Time Register): This register is incremented in 1-hour intervals once TCEN in TIMECON is active. The HOUR SFR counts from 0 to 23 before rolling over to 0.
SFR Address: A5H
Power-On Default: 00H
Bit Addressable: No
Valid Value: 0 to 23 decimal

(vii) Timers/Counters: Timers on a standard 8052 increment by 1 with each machine cycle. On the ADuC841/ADuC842/ADuC843, one machine cycle is equal to one clock cycle; therefore the timers increment at the same rate as the core clock.

The ADuC841/ADuC842/ADuC843 has three 16-bit timer/counters: Timer 0, Timer 1, and Timer 2. The timer/counter hardware is included on-chip to relieve the processor core of the overhead inherent in implementing timer/counter functionality in software. Each timer/counter consists of two 8-bit registers: THx and TLx (x = 0, 1, and 2). All three can be configured to operate either as timers or as event counters.

In timer function, the TLx register is incremented every machine cycle. Thus, one can think of it as counting machine cycles. Since a machine cycle on a single-cycle core consists of one core clock period, the maximum count rate is the core clock frequency.

In counter function, the TLx register is incremented by a 1-to-0 transition at its corresponding external input pin: T0, T1, or T2. When the samples show a high in one cycle and a low in the next cycle, the count is incremented. Since it takes two machine cycles (two core clock periods) to recognize a 1-to-0 transition, the maximum count rate is half the core clock frequency.

There are no restrictions on the duty cycle of the external input signal, but to ensure that a given level is sampled at least once before it changes, it must be held for a minimum of one full machine cycle. User configuration and control of all timer operating modes is achieved via three SFRs:

TMOD, TCON:  Control and configuration for Timers 0 and 1.

T2CON: Control and configuration for Timer 2.

a) **TMOD:** Timer/Counter 0 and 1 Mode Register
   SFR Address: 89H
   Power-On Default: 00H
   Bit Addressable: No

The TMOD SFR bit designations are given in table 3-10.

Table 3-10: TMOD SFR bit designations

| Bit No. | Name | Description |
|---------|------|-------------|
| 7 | Gate | Timer 1 Gating Control. |
| | | Set by software to enable Timer/Counter 1 only while the $\overline{INT1}$ pin is high and the TR1 control bit is set. |
| | | Cleared by software to enable Timer 1 whenever the TR1 control bit is set. |
| 6 | C/T | Timer 1 Timer or Counter Select Bit. |
| | | Set by software to select counter operation (input from T1 pin). |
| | | Cleared by software to select timer operation (input from internal system clock). |
| 5 | M1 | Timer 1 Mode Select Bit 1 (Used with M0 Bit). |
| 4 | M0 | Timer 1 Mode Select Bit 0. |

| M1 | M0 | |
|----|----|--|
| 0 | 0 | TH1 operates as an 8-bit timer/counter. TL1 serves as 5-bit prescaler. |
| 0 | 1 | 16-Bit Timer/Counter. TH1 and TL1 are cascaded; there is no prescaler. |
| 1 | 0 | 8-Bit Autoreload Timer/Counter. TH1 holds a value that is to be reloaded into TL1 each time it overflows. |
| 1 | 1 | Timer/Counter 1 Stopped. |

| Bit No. | Name | Description |
|---------|------|-------------|
| 3 | Gate | Timer 0 Gating Control. |
| | | Set by software to enable Timer/Counter 0 only while the $\overline{INT0}$ pin is high and the TR0 control bit is set. |
| | | Cleared by software to enable Timer 0 whenever the TR0 control bit is set. |
| 2 | C/T | Timer 0 Timer or Counter Select Bit. |
| | | Set by software to select counter operation (input from T0 pin). |
| | | Cleared by software to select timer operation (input from internal system clock). |
| 1 | M1 | Timer 0 Mode Select Bit 1. |
| 0 | M0 | Timer 0 Mode Select Bit 0. |

| M1 | M0 | |
|----|----|--|
| 0 | 0 | TH0 operates as an 8-bit timer/counter. TL0 serves as a 5-bit prescaler. |
| 0 | 1 | 16-Bit Timer/Counter. TH0 and TL0 are cascaded; there is no prescaler. |
| 1 | 0 | 8-Bit Autoreload Timer/Counter. TH0 holds a value that is to be reloaded into TL0 each time it overflows. |
| 1 | 1 | TL0 is an 8-bit timer/counter controlled by the standard Timer 0 control bits. |
| | | TH0 is an 8-bit timer only, controlled by Timer 1 control bits. |

b) **TCON:** Timer/Counter 0 and 1 Control Register

SFR Address: 88H

Power-On Default: 00H

Bit Addressable: Yes

The TCON SFR bit designations are given in table 3-11.

## Table 3-11: TCON SFR bit designations

| Bit No. | Name | Description |
|---|---|---|
| 7 | TF1 | Timer 1 Overflow Flag. |
| | | Set by hardware on a Timer/Counter 1 overflow. |
| | | Cleared by hardware when the program counter (PC) vectors to the interrupt service routine. |
| 6 | TR1 | Timer 1 Run Control Bit. |
| | | Set by the user to turn on Timer/Counter 1. |
| | | Cleared by the user to turn off Timer/Counter 1. |
| 5 | TF0 | Timer 0 Overflow Flag. |
| | | Set by hardware on a Timer/Counter 0 overflow. |
| | | Cleared by hardware when the PC vectors to the interrupt service routine. |
| 4 | TR0 | Timer 0 Run Control Bit. |
| | | Set by the user to turn on Timer/Counter 0. |
| | | Cleared by the user to turn off Timer/Counter 0. |
| 3 | IE1[1] | External Interrupt 1 ($\overline{INT1}$) Flag. |
| | | Set by hardware by a falling edge or by a zero level being applied to the external interrupt pin, $\overline{INT1}$, depending on the state of Bit IT1. |
| | | Cleared by hardware when the PC vectors to the interrupt service routine only if the interrupt was transition-activated. If level-activated, the external requesting source controls the request flag, rather than the on-chip hardware. |
| 2 | IT1[1] | External Interrupt 1 (IE1) Trigger Type. |
| | | Set by software to specify edge-sensitive detection, i.e., 1-to-0 transition. |
| | | Cleared by software to specify level-sensitive detection, i.e., zero level. |
| 1 | IE0[1] | External Interrupt 0 ($\overline{INT0}$) Flag. |
| | | Set by hardware by a falling edge or by a zero level being applied to external interrupt pin $\overline{INT0}$, depending on the state of Bit IT0. |
| | | Cleared by hardware when the PC vectors to the interrupt service routine only if the interrupt was transition-activated. If level-activated, the external requesting source controls the request flag, rather than the on-chip hardware. |
| 0 | IT0[1] | External Interrupt 0 (IE0) Trigger Type. |
| | | Set by software to specify edge-sensitive detection, i.e., 1-to-0 transition. |
| | | Cleared by software to specify level-sensitive detection, i.e., zero level. |

[1]These bits are not used in the control of Timer/Counter 0 and 1, but are used instead in the control and monitoring of the external $\overline{INT0}$ and $\overline{INT1}$ interrupt pins.

c) **Timer/Counter 0 and 1 Data Registers:** Each timer consists of two 8-bit registers. These can be used as independent registers or combined into a single 16-bit register depending on the timer mode configuration.

TH0 and TL0: Timer 0 high byte and low byte.
SFR Address = 8CH, 8AH respectively.

TH1 and TL1: Timer 1 high byte and low byte.
SFR Address = 8DH, 8BH respectively.

d) **T2CON: Timer/Counter 2 Control Register**
SFR Address: C8H

Power-On Default: 00H

Bit Addressable: Yes

The T2CON SFR bit designations are given in table 3-12.

Table 3-12: T2CON SFR bit designations

| Bit No. | Name | Description |
|---------|------|-------------|
| 7 | TF2 | Timer 2 Overflow Flag. |
| | | Set by hardware on a Timer 2 overflow. TF2 cannot be set when either RCLK = 1 or TCLK = 1. |
| | | Cleared by user software. |
| 6 | EXF2 | Timer 2 External Flag. |
| | | Set by hardware when either a capture or reload is caused by a negative transition on T2EX and EXEN2 = 1. |
| | | Cleared by user software. |
| 5 | RCLK | Receive Clock Enable Bit. |
| | | Set by the user to enable the serial port to use Timer 2 overflow pulses for its receive clock in serial port Modes 1 and 3. |
| | | Cleared by the user to enable Timer 1 overflow to be used for the receive clock. |
| 4 | TCLK | Transmit Clock Enable Bit. |
| | | Set by the user to enable the serial port to use Timer 2 overflow pulses for its transmit clock in serial port Modes 1 and 3. |
| | | Cleared by the user to enable Timer 1 overflow to be used for the transmit clock. |
| 3 | EXEN2 | Timer 2 External Enable Flag. |
| | | Set by the user to enable a capture or reload to occur as a result of a negative transition on T2EX if Timer 2 is not being used to clock the serial port. |
| | | Cleared by the user for Timer 2 to ignore events at T2EX. |
| 2 | TR2 | Timer 2 Start/Stop Control Bit. |
| | | Set by the user to start Timer 2. |
| | | Cleared by the user to stop Timer 2. |
| 1 | CNT2 | Timer 2 Timer or Counter Function Select Bit. |
| | | Set by the user to select counter function (input from external T2 pin). |
| | | Cleared by the user to select timer function (input from on-chip core clock). |
| 0 | CAP2 | Timer 2 Capture/Reload Select Bit. |
| | | Set by the user to enable captures on negative transitions at T2EX if EXEN2 = 1. |
| | | Cleared by the user to enable autoreloads with Timer 2 overflows or negative transitions at T2EX when EXEN2 = 1. When either RCLK = 1 or TCLK = 1, this bit is ignored and the timer is forced to autoreload on Timer 2 overflow. |

e) **Timer/Counter2 Data Registers:** Timer/Counter 2 also has two pairs of 8-bit data registers associated with it. These are used as both timer data registers and as timer capture/reload registers.

**TH2 and TL2**

Timer 2: Data high byte and low byte.

SFR Address: CDH, CCH respectively.

**RCAP2H and RCAP2L**

Timer 2: Capture/reload high and low byte.

SFR Address: CBH, CAH respectively.

64

**3.1.5 OLED display:** Micro-OLED is manufactured by 4D systems. An Organic Light Emitting Diode (OLED) is a Light-Emitting Diode (LED) in which the emissive electroluminescent layer is a film of organic compounds which emit light in response to an electric current. This layer of organic semiconductor material is situated between two electrodes. Generally, at least one of these electrodes is transparent. OLEDs are used in television screens, computer monitors, small, portable system screens such as mobile phones and PDAs, watches, advertising, information, and indication. OLEDs are also used in light sources for space illumination and in large-area light-emitting elements. Due to their early stage of development, they typically emit less light per unit area than inorganic solid-state based LED point-light sources.

An OLED display functions without a backlight. Thus, it can display deep black levels and can be thinner and lighter than liquid crystal displays. In low ambient light conditions such as dark rooms, an OLED screen can achieve a higher contrast ratio than an LCD using either cold cathode fluorescent lamps or the more recently developed LED backlight.

There are two main families of OLEDs: those based upon small molecules and those employing polymers. Adding mobile ions to an OLED creates a Light-emitting Electrochemical Cell or LEC, which has a slightly different mode of operation.

OLED displays can use either passive-matrix (PMOLED) or active-matrix addressing schemes. Active-matrix OLEDs (AMOLED) require a thin-film transistor backplane to switch each individual pixel on or off, and can make higher resolution and larger size displays possible.

Some of the features of OLED include:

1. Low thermal dissipation
2. High pixel density
3. Excellent clarity
4. Thinner and lighter than LCDs

The μOLED-3202X-P1 series are a compact and cost effective all in one 'SMART' display modules using the latest state of the art Active Matrix OLED (AMOLED) technology with an embedded PICASO-GFX graphics controller that delivers 'stand-alone' functionality to any project. Figure 3-22 shows the μOLED-3202X-P1.

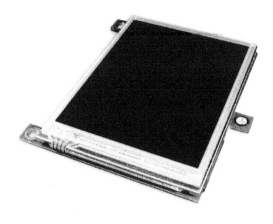

Figure 3-22: Micro-OLED display

The µOLED-3202X-P1 series of modules are aimed at being integrated into a variety of different applications via a wealth of features designed to facilitate any given functionality quickly and cost effectively and thus reduce 'time to market'. These features are as follows:

There are 4 modules in the µOLED-3202X-P1 series:

1. µOLED-32024-P1

   Diagonal: 2.4"

   Screen Outline: 42.0 x 52.6 mm

   Active Area: 36.7 x 49.0 mm

2. µOLED-32024-P1T

   Same as µOLED-32024-P1 but with resistive touch screen.

3. µOLED-32028-P1

   Diagonal: 2.83"

   Screen Outline: 49.1 x 67.3 mm

   Active Area: 43.2 x 57.6 mm

4. µOLED-32028-P1T

   Same as µOLED-32028-P1 but with resistive touch screen.

I have used µOLED-32028-P1. Some common features for all these four modules are:

1. QVGA 240 x RGB x 320 pixel resolution with 256, 65K or 262K true to life colours enhanced AMOLED screen.

2. Near 180 degree viewing angle.

3. All modules use the same controller board. PCB Size: 49.1 x 67.3 x 11.0mm.

4. Easy 5 pin user interface ($V_{CC}$, $T_X$, $R_X$, GND, RESET) to any 4D micro-USB module such as the μUSB-MB5 or the μUSB-CE5.

5. Voltage supply from 4.5V to 5.5V, current @ 90mA nominal when using a 5.0V supply source.

6. Onboard micro-SD (μSD) memory card adaptor with full FAT16 file support for storing and executing 4DGL programs, files, icons, images, animations, video clips and audio wave files. 64Mb to 2Gig μSD memory cards can be purchased separately.

7. Powered by the fully integrated PICASO-GFX Graphics Processor (PICASO-GFX chip is also available for OEM volume users).

8. 2 x 30 pin headers for I/O expansion and future plug-in daughter boards.

9. Audio amplifier with a tiny 8 ohms speaker for sound generation and wave file playback.

10. Mechanical support via mounting tabs which can be snapped off.

Figure 3-23 shows the circuit diagram of μOLED-3202X-P1 powered by PICASO-GFX.

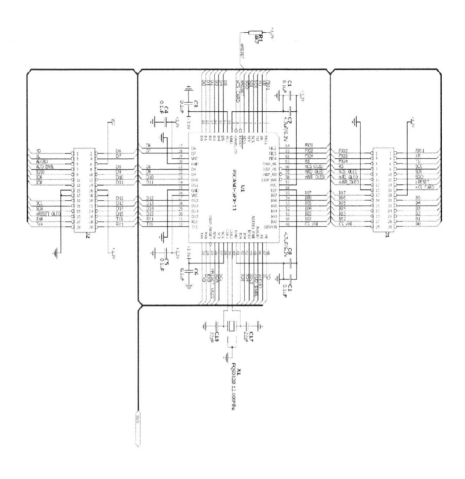

Figure 3-23: Circuit diagram of μOLED-3202X-P1 powered by PICASO-GFX

Figure 3-24 shows the pin diagram of μOLED-3202X-P1.

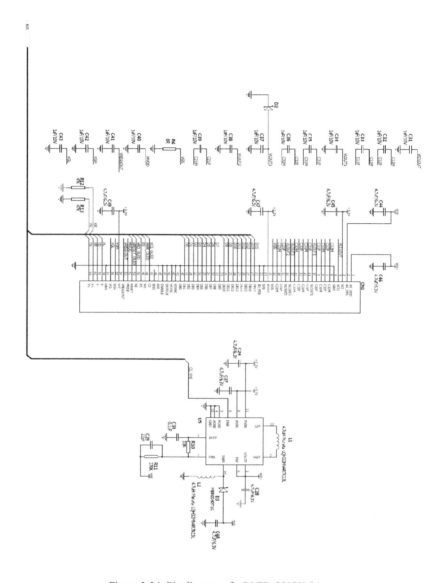

Figure 3-24: Pin diagram of µOLED-3202X-P1

Figure 3-25 shows the micro-USB interface with the USD card.

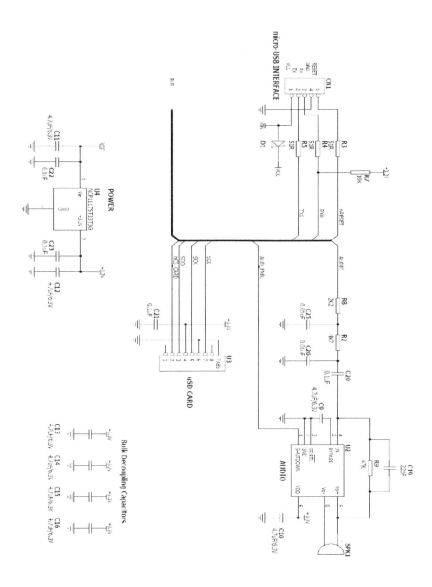

Figure 3-25: Micro-USB interface with the USD card

Table 3-13 shows the pin description of power, serial and micro-USB interface.

Table 3-13: Pin description of power, serial and micro-USB interface

| Power, Serial and micro-USB Interface | | |
|---|---|---|
| Pin | Function | Description |
| 1 | VIN | Main Power Supply input 4.5Volts to 5.5Volts. Nominal @ 5Volts. |
| 2 | TX | Serial Transmit Pin (Data Out), COM0 TX. CMOS levels 0V to 3.3V |
| 3 | RX | Serial Receive Pin (Data In), COM0 RX. CMOS levels 0V to VIN. |
| 4 | GND | Ground. |
| 5 | RES | External RESET signal for the module and PICASO chip. Pull this pin Low for 20μsec or longer to Reset the module. Not required for normal usage. |

### 3.1.5.1 Expansion Ports Pin Description

Table 3-14 and 3-15 shows the pin description of Expansion Ports J1 and J2 respectively.

Table 3-14: Pin description of Expansion Port J1

| EXPANSION PORT J1 (for future 4D add-on modules) | | |
|---|---|---|
| Pin | Label | Description |
| 1 | FIO2 | Factory IO2 pin. (Reserved, do not use). |
| 2 | FIO1 | Factory IO1 pin. (Reserved, do not use). |
| 3 | FIO3 | Factory IO3 pin. (Reserved, do not use). |
| 4 | XR | 4-Wire resistive touch screen right signal. (Reserved, do not use). |
| 5 | FIO4 | Factory IO4 pin. (Reserved, do not use). |
| 6 | YU | 4-Wire resistive touch screen top signal. (Reserved, do not use). |
| 7 | RS | Display register select signal. (Reserved, do not use). |
| 8 | SCK | SPI serial clock output for external SD card use only. |
| 9 | nCS_OLED | OLED chip select signal. (Reserved, do not use). |
| 10 | SDI | SPI serial data input for external SD card use only. |
| 11 | nRD_OLED | OLED read strobe signal. (Reserved, do not use). |
| 12 | SDO | SPI serial data output for external SD card use only. |
| 13 | nWR_OLED | OLED write strobe signal. (Reserved, do not use). |
| 14 | nRESET | Master RESET. Pull this pin Low for 20µsec or longer to Reset the module. |
| 15 | GND | Ground. |
| 16 | nCS_CARD | SD memory card chip select for external SD card use only. |
| 17 | IO7 | General Purpose Input Output 7 pin. |
| 18 | 3.3V | Regulated 3.3 Volts output, available current max 400mA. |
| 19 | IO6 | General Purpose Input Output 6 pin. |
| 20 | D5 | OLED data bus bit 5. (Reserved, do not use). |
| 21 | IO5 | General Purpose Input Output 5 pin. |
| 22 | D4 | OLED data bus bit 4. (Reserved, do not use). |
| 23 | IO4 | General Purpose Input Output 4 pin. |
| 24 | D3 | OLED data bus bit 3. (Reserved, do not use). |
| 25 | IO3 | General Purpose Input Output 3 pin. |
| 26 | D2 | OLED data bus bit 2. (Reserved, do not use). |
| 27 | IO2 | General Purpose Input Output 2 pin. |
| 28 | D1 | OLED data bus bit 1. (Reserved, do not use). |
| 29 | CS_VHI | OLED DC-DC circuit enable signal. (Reserved, do not use). |
| 30 | D0 | OLED data bus bit 0. (Reserved, do not use). |

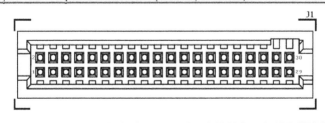

Table 3-15: Pin description of Expansion Port J2

| EXPANSION PORT J2 (for future 4D add-on modules) | | |
|---|---|---|
| Pin | Label | Description |
| 1 | YD | 4-Wire resistive touch screen bottom signal. (Reserved, do not use). |
| 2 | D6 | OLED data bus bit 6. (Reserved, do not use). |
| 3 | XL | 4-Wire resistive touch screen left signal. (Reserved, do not use). |
| 4 | D7 | OLED data bus bit 7. (Reserved, do not use). |
| 5 | AUDIO | Pulse width modulated Audio output from PICASO. This pin is also input to the onboard audio amplifier. |
| 6 | VCC | Main Power Supply input 4.5Volts to 5.5Volts. Nominal @ 5Volts. |
| 7 | AUDIO_ENBL | Logic Low will enable the audio amplifier, logic High will disable it. |
| 8 | D8 | OLED data bus bit 8. (Reserved, do not use). |
| 9 | IO10 | General Purpose Input Output 10 pin. |
| 10 | D9 | OLED data bus bit 9. (Reserved, do not use). |
| 11 | IO9 | General Purpose Input Output 9 pin. |
| 12 | D10 | OLED data bus bit 10. (Reserved, do not use). |
| 13 | IO8 | General Purpose Input Output 8 pin. |
| 14 | D11 | OLED data bus bit 11. (Reserved, do not use). |
| 15 | GND | Ground. |
| 16 | 3.3V | Regulated 3.3 Volts output, available current max 400mA. |
| 17 | GND | Ground. |
| 18 | 3.3V | Regulated 3.3 Volts output, available current max 400mA. |
| 19 | N.C. | No Connect. |
| 20 | D12 | OLED data bus bit 12. (Reserved, do not use). |
| 21 | SCL | I2C clock output. |
| 22 | D13 | OLED data bus bit 13. (Reserved, do not use). |
| 23 | SDA | I2C bi-directional data. |
| 24 | D14 | OLED data bus bit 14. (Reserved, do not use). |
| 25 | nRESET_OLED | OLED Reset signal. (Reserved, do not use). |
| 26 | D15 | OLED data bus bit 15. (Reserved, do not use). |
| 27 | RX0 | Asynchronous serial port 0 receive pin. COM0 Rx. |
| 28 | TX1 | Asynchronous serial port 1 transmit pin. COM1 Tx. |
| 29 | TX0 | Asynchronous serial port 0 transmit pin. COM0 Tx. |
| 30 | RX1 | Asynchronous serial port 1 receive pin. COM1 Rx. |

### 3.1.5.2 USB to Serial Interface – Micro USB

The μOLED-3202X-P1 module is required to be interfaced to a PC for uploading the PICASO-GFX chip. Using a standard USB cable and any one of the micro-USB modules (μUSB-MB5 or μUSB-CE5) as shown in figure 3-26, a PC to μOLED-3202X-P1 connection can be achieved simply.

The micro-USB interface is also used for PmmC (Personality module micro Code) uploads. The PmmC allows the latest Operating System and 4DVM (4D Virtual Machine) upgrades for the PICASO-GFX chip. The micro-USB module simply connects to the μOLED-3202X-P1 module and captures the USB data and converts it into serial CMOS level (0 to 3.3V) data.

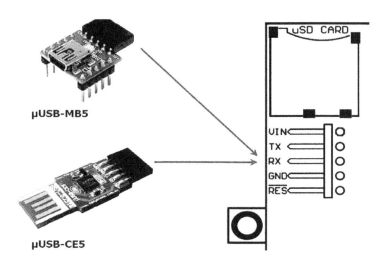

Figure 3-26: USB to serial interface

# CHAPTER 4

# HARDWARE IMPLEMENTATION

## 4.1 Circuit diagram

The circuit diagram of pulse oximeter consists of the following stages:

1. PPG probe
2. Switching and LED driver circuit
3. I/V converter
4. Amplification stage
5. ADC stage
6. ADuC842 microcontroller
7. LCD/OLED display

The circuit diagram for the pulse oximeter is shown in figure 4-1.

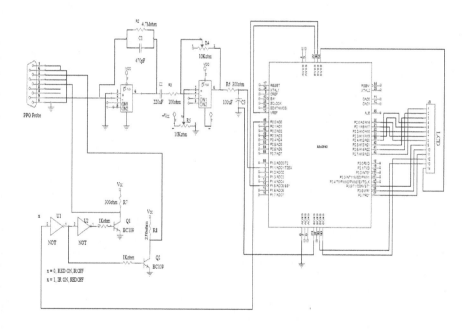

Figure 4-1: Circuit diagram of pulse oximeter

This circuit has been made in ORCAD version 9.1 The ORCAD capture of the circuit diagram is shown in figure 4-2.

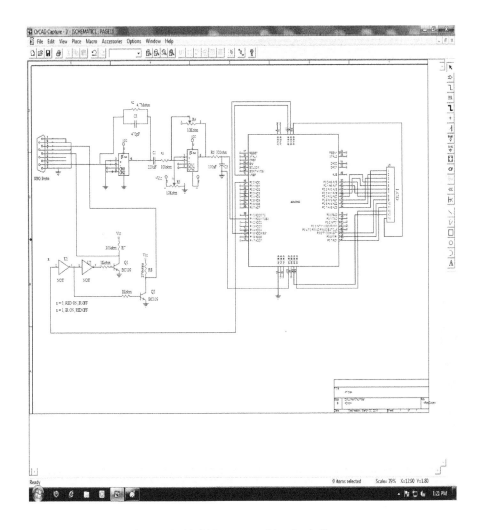

Figure 4-2: ORCAD capture of the circuit diagram

## 4.2 Description of the circuit

The DB9 connector is connected with the PPG probe. Pin no. 1, 5, 6 and 7 of DB9 connector are grounded. Pin no. 4 and 8 are not connected. Pin no. 2 and 3 are used with the LEDs. Pin no. 9 is the output from the DB9 connector which is connected to pin no.2 of the IC UA741 which is behaving as the I/V converter.

The switching and LED driver circuit consist of two transistors Q1 and Q2 (BC109). There are two NOT gates in the circuit. When one transistor is ON, other will be OFF and vice-versa. IC 7404 is used as a NOT gate. Figure 4-3 shows the pin diagram of IC 7404.

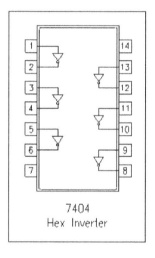

Figure 4-3: Pin diagram of IC 7404

Here only two NOT gates are used in the circuit and rest of the NOT gates are not used. The pin description of IC 7404 is given in table 4-1. The dimension drawing of IC 7404 is given in figure 4-4. Table 4-2 gives the technical data of IC 7404.

Table 4-1: Pin description of IC 7404

| Pin Description of IC 7404 | |
|---|---|
| **Pin Number** | **Description** |
| 1 | A Input Gate 1 |
| 2 | Y Output Gate 1 |
| 3 | A Input Gate 2 |
| 4 | Y Output Gate 2 |
| 5 | A Input Gate 3 |
| 6 | Y Output Gate 3 |
| 7 | Ground |
| 8 | Y Output Gate 4 |
| 9 | A Input Gate 4 |
| 10 | Y Output Gate 5 |
| 11 | A Input Gate 5 |
| 12 | Y Output Gate 6 |
| 13 | A Input Gate 6 |
| 14 | Positive Supply |

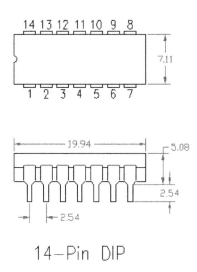

14-Pin DIP

Figure 4-4: Dimension drawing of IC 7404

Table 4-2: Technical data of IC 7404

| Technical Data |
|---|

**Absolute Maximum Ratings**

| | |
|---|---|
| Supply Voltage | 7V |
| Input Voltage | 5.5V |
| Operating Free Air Temperature | 0oC to +70oC |
| Storage Temperature Range | -65oC to +150oC |

**Recommended Operating Conditions**

| Symbol | Parameter | Min | Nom | Max | Units |
|---|---|---|---|---|---|
| Vcc | Supply Voltage | 4.75 | 5 | 5.25 | V |
| Vih | HIGH Level Input Voltage | 2 | | | V |
| Vil | LOW Level Input Voltage | | | 0.8 | V |
| Ioh | HIGH Level Output Current | | | -0.4 | mA |
| Iol | LOW Level Output Current | | | 16 | mA |
| Ta | Free Air Operating Temperature | 0 | | 70 | oC |

**Electrical Characteristics**

| Symbol | Parameter | Conditions | Min | Typ | Max | Units |
|---|---|---|---|---|---|---|
| Vi | Input Clamp Voltage | Vcc=Min,Ii=-12mA | | | -1.5 | V |
| Voh | HIGH Level Output Voltage | Vcc=Min,Ioh=MAX,Vil=MAX | 2.4 | 3.4 | | V |
| Vol | LOW Level Output Voltage | Vcc=Min,Iol=MAX,Vih=MAX | | 0.2 | 0.4 | V |
| Ii | Input Current@MAX Input Voltage | Vcc=Max,Vi=5.5V | | | 1 | mA |
| Iih | HIGH Level Input Current | Vcc=Max,Vi=2.4V | | | 40 | uA |
| Iil | LOW Level Input Current | Vcc=Max,Vi=0.4V | | | -1.6 | mA |
| Ios | Short Circuit Output Current | Vcc=Max | -18 | | -55 | mA |
| Icch | Supply Current with Outputs HIGH | Vcc=Max | | 4 | 8 | mA |
| Iccl | Supply Current with Outputs LOW | Vcc=Max | | 12 | 22 | mA |

| Switching Characteristics at $V_{cc}$=5V,Ta=25oC | | | | | |
|---|---|---|---|---|---|
| Symbol | Parameter | Conditions | Min | Max | Units |
| tplh | Propogation Delay Time LOW-to-HIGH Level Output | Cl=15pF Rl=400R | | 22 | nS |
| tphl | Propogation Delay Time HIGH-to-LOW Level Output | | | 15 | nS |

IC UA741 is used as current to voltage converter and for the amplification of the signal. A low-pass RC filter is employed to eliminate the noise completely.

The circuit was first implemented on the bread board and then implemented on the PCB. Figures 4-5 and 4-6 show the bread board and PCB implementation of the circuit respectively.

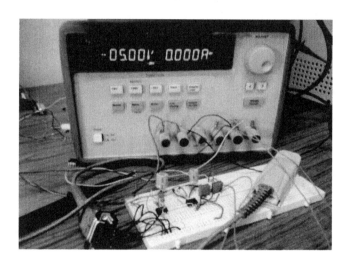

Figure 4-5: Bread board testing of the circuit

Figure 4-6: PCB implementation

Now, ADuC842 microcontroller is interfaced with PICASO-GFX2 microcontroller and the display to be shown on micro-OLED display.

The final circuit of the pulse oximeter implemented on PCB is shown in figure 4-7. Figure 4-8 shows the connections made with the ADuC842 microcontroller.

Figure 4-7: Final circuit implemented on PCB

Figure 4-8: Connections with the ADuC842 microcontroller

**4.2.1 About PICASO-GFX2 microcontroller:** The PICASO-GFX2 is a custom embedded 4DGL graphics controller designed to interface with many popular OLED and LCD display panels. Powerful graphics, text, image, animation and countless more features are built right inside the chip. It offers a simple plug-n-play interface to many 16-bit 80-Series colour LCD and OLED displays. Figure 4-9 shows PICASO-GFX2 microcontroller from 4D labs.

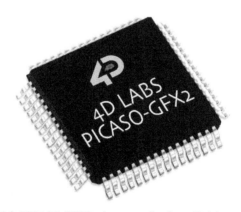

Figure 4-9: PICASO-GFX2 microcontroller from 4D labs

The chip is designed to work with minimal design effort and all of the data and control signals are provided by the chip to interface directly to the display. The PICASO-GFX2 belongs to a family of processors powered by a highly optimised soft core virtual engine, E.V.E. (Extensible Virtual

83

Engine). EVE is a proprietary, high performance virtual processor with an extensive byte-code instruction set.

The device offers modest but comprehensive I/O features and can interface to I2C, serial, digital, buttons, joystick and many more. A basic system font is included, and unlimited customisable fonts with fixed or proportional spacing can be created using the free FONT-Tool provided.

All of the display built-in driver libraries implement and share the same high-level function interface. This allows your GUI application to be portable to different display controller types.

In short, the PICASO-GFX2 offers one of the most flexible embedded graphics solutions available.

Some of the features of PICASO-GFX2 are:

1. Low-cost OLED, LCD and TFT display graphics user interface solution.
2. Ideal as a standalone embedded graphics processor or interface to any host controller as a graphics co-processor.
3. Connect to any colour display that supports an 80-Series 16 bit wide CPU interface. All data and control signals are provided.
4. Built in high performance virtual processor core (EVE) with an extensive byte-code instruction set optimised for 4DGL, the high level 4D Graphics Language.
5. Comprehensive set of built in graphics and multimedia services.
6. Display full colour images, animations, icons and video clips.
7. 15K bytes of flash memory for user code storage and 14K bytes of SRAM for user variables.
8. 13 Digital I/O pins.
9. I2C interface (Master).
10. D0….D15, RD, WR, RS, CS – Display interface
11. FAT16 file services.
12. Asynchronous hardware serial ports with Auto-Baud feature.
13. SPI interface support for SDHC/SD memory card for multimedia storage and data logging purposes (μSD with up to 2GB and SDHC memory cards starting from 4GB and above).
14. 4-Wire resistive touch panel interface.
15. Audio support for wave files and complex sound generation with a dedicated 16-bit PWM audio output.
16. 8 x 16 bit timers with 1 millisecond resolution.
17. Single 3.3 Volt Supply @25mA typical.
18. Available in a 64 pin TQFP 10mm x 10mm package.
19. RoHS compliant.

Figure 4-10 shows the PICASO-GFX2 microcontroller which is in-built on the back panel of the OLED module.

Figure 4-10: PICASO-GFX2 microcontroller which is in-built on the back panel of the OLED module

Figure 4-11 shows the internal block diagram of PICASO-GFX2 microcontroller and figure 4-12 shows the pin diagram of PICASO-GFX2 microcontroller.

Table 4-3 gives the pin description of PICASO-GFX2 microcontroller.

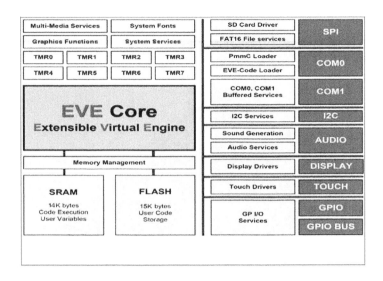

Figure 4-11: PICASO-GFX2 internal block diagram

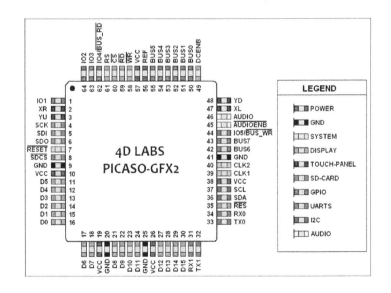

Figure 4-12: Pin diagram of PICASO-GFX2 microcontroller

Table 4-3: Pin description of PICASO-GFX2 microcontroller

| Pin | Symbol | I/O | Description |
|-----|--------|-----|-------------|
| 1 | IO1 | I/O | General Purpose IO1 pin. This pin is 5.0V tolerant. |
| 2 | XR | A | 4-Wire Resistive Touch Screen Right signal. Connect this pin to XR or X+ signal of the touch panel. |
| 3 | YU | A | 4-Wire Resistive Touch Screen Up signal. Connect this pin to YU or Y+ signal of the touch panel. |
| 4 | SCK | O | SPI Serial Clock output. SD memory card use only. Connect this pin to the SPI Serial Clock (SCK) signal of the memory card. |
| 5 | SDI | I | SPI Serial Data Input. SD memory card use only. Connect this pin to the SPI Serial Data Out (SDO) signal of the memory card. |
| 6 | SDO | O | SPI Serial Data Output. SD memory card use only. Connect this pin to the SPI Serial Data In (SDI) signal of the memory card. |
| 7 | RESET | I | Master Reset signal. Connect a 4.7K resistor from this pin to VCC. |
| 8 | SDCS | O | SD Memory-Card Chip Select. SD memory card use only. Connect this pin to the Chip Enable (CS) signal of the memory card. |

| Pin | Symbol | I/O | Description |
|---|---|---|---|
| 9, 20, 25, 41 | GND | P | Device Ground. |
| 10, 19, 26, 38, 57 | VCC | P | Device Positive Supply. |
| 11 | D5 | I/O | Display Data Bus bit 5. |
| 12 | D4 | I/O | Display Data Bus bit 4. |
| 13 | D3 | I/O | Display Data Bus bit 3. |
| 14 | D2 | I/O | Display Data Bus bit 2. |
| 15 | D1 | I/O | Display Data Bus bit1. |
| 16 | D0 | I/O | Display Data Bus bit 0. |
| 17 | D6 | I/O | Display Data Bus bit 6. |
| 18 | D7 | I/O | Display Data Bus bit 7. |
| 21 | D8 | I/O | Display Data Bus bit 8. |
| 22 | D9 | I/O | Display Data Bus bit 9. |
| 23 | D10 | I/O | Display Data Bus bit 10. |
| 24 | D11 | I/O | Display Data Bus bit 11. |
| 27 | D12 | I/O | Display Data Bus bit 12. |
| 28 | D13 | I/O | Display Data Bus bit 13. |
| 29 | D14 | I/O | Display Data Bus bit 14. |
| 30 | D15 | I/O | Display Data Bus bit 15. |
| 31 | RX1 | I/O | General Purpose I/O Port, bit 0. This pin is 5.0V tolerant. |
| 32 | TX1 | I/O | General Purpose I/O Port, bit 1. This pin is 5.0V tolerant. |
| 33 | TX0 | O | Asynchronous Serial port Transmit pin, TX. Connect this pin to host micro-controller Serial Receive (Rx) signal. The host receives data from PICASO-GFX2 via this pin. This pin is 5.0V tolerant. |
| 34 | RX0 | I | Asynchronous Serial port Receive pin, RX. Connect this pin to host micro-controller Serial Transmit (Tx) signal. The host transmits data to PICASO-GFX2 via this pin. This pin is 5.0V tolerant. |
| 35 | RES | O | Display RESET. PICASO-GFX2 initialises the display by strobing this pin LOW. Connect this pin to the Reset (RES) signal of the display. |
| 36 | SDA | I/O | I2C Data In/Out. |
| 37 | SCL | O | I2C Clock Output. |
| 39 | CLK1 | I | Device Clock input 1 of a 12Mhz crystal. |
| 40 | CLK2 | O | Device Clock input 2 of a 12Mhz crystal. |
| 42 | BUS6 | I/O | General Purpose Parallel I/O BUS(0..7), bit 6. This pin is 5.0V tolerant. |
| 43 | BUS7 | I/O | General Purpose Parallel I/O BUS(0..7), bit 7. This pin is 5.0V tolerant. |
| 44 | IO5/BUS_WR | I/O | General Purpose IO5 pin. Also used for BUS_WR signal to write and latch the data to the parallel GPIO BUS(0..7). |
| 45 | AUDENB | O | Audio Enable. Connect this pin to amplifier control. LOW: Enable external Audio amplifier. HIGH : Disable external Audio amplifier. |
| 46 | AUDIO | O | Pulse Width Modulated (PWM) Audio output. Connect this pin to a 2 |

| Pin | Symbol | I/O | Description |
|---|---|---|---|
| | | | stage low pass filter then into an audio amplifier. |
| 47 | XL | O | 4-Wire Resistive Touch Screen Left signal. Connect this pin to XL or X- signal of the touch panel. |
| 48 | YD | O | 4-Wire resistive touch screen bottom signal. Connect this pin to YD or Y- signal of the touch panel. |
| 49 | DCENB | O | DC-DC high voltage enable signal. This maybe the high voltage that drives the LCD backlight or the OLED panel supply. **High**: Enable DC-DC converter. **Low** : Disable DC-DC converter. |
| 50 | BUS0 | I/O | General Purpose Parallel I/O BUS(0..7), bit 0. This pin is 5.0V tolerant. |
| 51 | BUS1 | I/O | General Purpose Parallel I/O BUS(0..7), bit 1. This pin is 5.0V tolerant. |
| 52 | BUS2 | I/O | General Purpose Parallel I/O BUS(0..7), bit 2. This pin is 5.0V tolerant. |
| 53 | BUS3 | I/O | General Purpose Parallel I/O BUS(0..7), bit 3. This pin is 5.0V tolerant. |
| 54 | BUS4 | I/O | General Purpose Parallel I/O BUS(0..7), bit 4. This pin is 5.0V tolerant. |
| 55 | BUS5 | I/O | General Purpose Parallel I/O BUS(0..7), bit 5. This pin is 5.0V tolerant. |
| 56 | REF | P | Internal voltage regulator filter capacitor. Connect a 4.7uF to 10uF capacitor from this pin to Ground. |
| 58 | WR | O | Display Write strobe signal. PICASO-GFX2 asserts this signal LOW when writing data to the display. Connect this pin to the Write (WR) signal of the display. |
| 59 | RD | | Display Read strobe signal. PICASO-GFX2 asserts this signal LOW when reading data from the display. Connect this pin to the Read (RD) signal of the display. |
| 60 | CS | O | Display Chip Select. PICASO-GFX2 asserts this signal LOW when accessing the display. Connect this pin to the Chip Select (CS) signal of the display. |
| 61 | RS | O | Display Register Select. **LOW**: Display index or status register is selected. **HIGH**: Display GRAM or register data is selected. Connect this pin to the Register Select (RS or A0 or C/D or similar naming convention) signal of the display. |
| 62 | IO4/BUS_RD | I/O | General Purpose IO4 pin. Also used for BUS_RD signal to read and latch the data in to the parallel GPIO BUS(0..7). |
| 63 | IO3 | I/O | General Purpose IO3 pin. This pin is 5.0V tolerant. |
| 64 | IO2 | I/O | General Purpose IO2 pin. This pin is 5.0V tolerant. |

I: Input, O: Output, A: Analogue, P: Power

Figure 4-13 illustrates how the PICASO-GFX2 internal memory is organised.

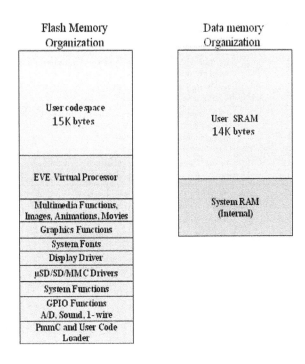

Figure 4-13: PICASO-GFX2 internal memory organization

Some of the applications of PICASO-GFX2 microcontroller are:

(a) General purpose embedded graphics.

(b) Elevator control systems.

(c) Point of sale terminals.

(d) Electronic gauges and metres.

(e) Test and measurement and general purpose instrumentation.

(f) Industrial control and Robotics.

(g) Automotive system displays.

(h) GPS navigation systems.

(i) Medical Instruments and applications.

(j) Home appliances and Smart Home Automation.

(k) Security and Access control systems.

(l) Gaming equipment.

(m) Aviation systems.

(n) HMI with touch panels.

Table 4-4 illustrates the operating conditions and table 4-5 illustrates the global characteristics based on operating conditions of PICASO-GFX2 microcontroller.

Table 4-4: Operating conditions of PICASO-GFX2 microcontroller

| Parameter | Conditions | Min | Typ | Max | Units |
|---|---|---|---|---|---|
| Supply Voltage (VCC) | | 3.0 | 3.3 | 3.6 | V |
| Operating Temperature | | -40 | -- | +80 | °C |
| External Crystal (Xtal) | | -- | 12.00 | -- | Mhz |
| Input Low Voltage (VIL) | VCC = 3.3V, all pins | VGND | -- | 0.2VCC | V |
| Input High Voltage (VIH) | VCC = 3.3V, non 5V tolerant pins | 0.8VCC | -- | VCC | V |
| Input High Voltage (VIH) | All GPIO pins, RX0 and TX0 pins | 0.8VCC | -- | 5.5 | V |

Table 4-5: Global characteristics based on operating conditions of PICASO-GFX2 microcontroller

| Parameter | Conditions | Min | Typ | Max | Units |
|---|---|---|---|---|---|
| Supply Current (ICC) | VCC = 3.3V | -- | 50 | 90 | mA |
| Internal Operating Frequency | Xtal = 12.00Mhz | -- | 48.00 | -- | Mhz |
| Output Low Voltage (VOL) | VCC = 3.3V, IOL = 3.4mA | -- | -- | 0.4 | V |
| Output High Voltage (VOH) | VCC = 3.3V, IOL = -2.0mA | 2.4 | -- | -- | V |
| A/D Converter Resolution | XR, YU  pins | 8 | -- | 10 | bits |
| Capacitive Loading | CLK1, CLK2 pins | -- | -- | 15 | pF |
| Capacitive Loading | All other pins | -- | -- | 50 | pF |
| Flash Memory Endurance | PmmC Programming | -- | 1000 | -- | E/W |

### 4.3 Testing and troubleshooting of the circuit

After testing the circuit on the bread board and implementing it on the PCB; next step was testing and troubleshooting of the circuit. I tested the circuit with the help of multimeter for various short circuited connections and checked the continuity of the circuit. Having found some of the short circuited connections, I troubleshoot the circuit.

### 4.4 Obtaining the waveform for the heart beat

For obtaining the waveform for the heart beat, we have to set the gain for both the potentiometers such that the signal is above ground level and no clipping of the waveform is there in both the positive as well as the negative half of the cycle and that too on the 5V scale. We have to vary the potentiometer so that the signal does not get attenuated and it should be above ground level also.

Gain of an amplifier is defined as the number of times the signal is boosted or raised in terms of amplitude or power. In other words we can say that the gain of an amplifier is the ratio of output to the input power or amplitude, and is usually measured in dB.

For 10K potentiometer which is connected to pin no.3 of second op-amp, we have to keep voltage=0V because it is offset voltage.

So by moving the slider of both the potentiometers and checking that the waveform does not get clipped for both positive and the negative half cycles, we obtain the following values:

For potentiometer which is connected to pin no.2, resistance value= 8.06 Kohm

For potentiometer which is connected to pin no.3, resistance value= 226 ohm

Voltage at pin no.3 = 0.1V

The waveform which we got on the DSO after moving the slider for both the potentiometers is as shown in figure 4-14.

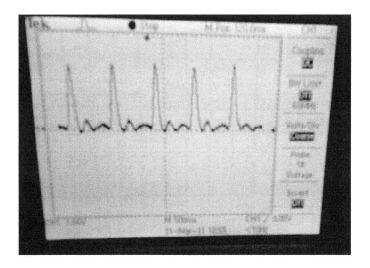

Figure 4-14: Waveform for the heart beat

The waveform which we obtained is inverted as the op-amps are used in inverting stage. From this waveform, we have to calculate heart beat (no. of beats per minute) and display the result on the LCD.

# CHAPTER 5

## SOFTWARE IMPLEMENTATION

### 5.1 Programming environment

#### 5.1.1 About the software

The software used is Aspire version 1.05. Aspire from Accutron Ltd. is a powerful Integrated Development Environment intended to ease cross-platform applications development, supporting both C and Assembly language development.

The aspire feature set includes:

1) Enhanced interactive interface for the development of an application including the support to projects, multiple windows interface with dockable windows, sophisticated text editor.

2) Working with multiple projects in a single workspace.

3) Creation, development and debug of an application in a single development cycle.

4) Editor supports working on multiple projects simultaneously, with multi-language advanced editing and syntax highlighting.

5) Ability to develop cross-platform applications using range of compilers and different target hardware.

6) Supports software simulation, hardware emulators, and in-circuit debuggers.

7) Convenient windows for viewing and altering target data.

8) New compilers are effortlessly plugged-in.

9) New target debugging engines (simulators, emulators, etc) can be merely plugged-in.

10) Drag and Drop intuitive interface.

#### 5.1.2 Compiler and assembler configuration

When the aspire is launched for the first time, the user will be asked to configure the Keil compiler and assembler for use with aspire as shown in figure 5-1.

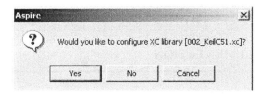

Figure 5-1: Configuring the Keil compiler and assembler

Click yes, if you have Keil installed and wish to configure aspire to use it. Click no if you do not wish configure. Click Cancel if you do not wish to configure at this time, but would like to be asked again the next time aspire is started.

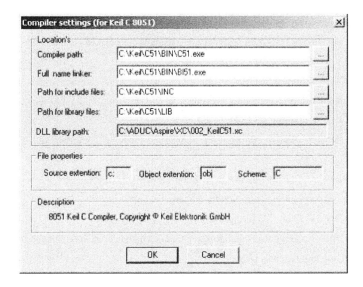

Figure 5-2: Compiler settings

Enter the path, the required files and press ok as shown in figure 5-2. The compiler configuration process will continue until the user has been asked whether or not to configure all available compilers.

### 5.1.3 Managing projects

#### 5.1.3.1 Creating a new project

On invoking aspire, we must create new or open an existing project. To create a new project, select New Project command from the Project menu as shown in figure 5-3 or right-click the workspace window and select New Project in the context menu.

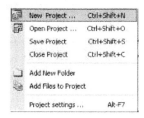

Figure 5-3: Project menu showing 'New Project' command option

Aspire displays the New Project dialog box first, so we can name our project and select the default compiler. Next the project settings dialogue will be presented. Project options can be revised at any time by selecting the Project settings command from the Project menu.

A new project is automatically added to the Workspace, made active, and supplied with Header, Info and Source workspace folders. We can add and remove files within the project, group our project files to these folders, or create and remove project folders. Use context menu to manage the project files and folders by right-clicking them on the workspace window.

To save a project using its existing name and folder, use the Save Project command. We can also create a new project by right-clicking on a project or a free space in the workspace window and selecting new project in the context menu.

Shortcuts:

Toolbar: No

Keys: CTRL+SHIFT+N

#### 5.1.3.2 Opening an existing project

To open an existing project, select Open Project command from the Project menu as shown in figure 5-4 or right-click the workspace window and select Open Project in the context menu.

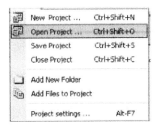

Figure 5-4: Project menu showing 'Open Project' command option

The Open Project dialog box is displayed by aspire, so we can choose a project. To close an opened project with its existing name and directory, use the Close Project command. We can also open a project by right-clicking on a project or a free space in the workspace window and selecting Open project in the context menu. Any opened project is added to the workspace, and made active. This means we can immediately work with it. At any time we can switch to any other project present in the workspace, or remove a project from the workspace.

Shortcuts:

Toolbar: No

Keys: CTRL+SHIFT+O

**5.1.3.3 Saving a project**

Use this command to save the active project to its current name and folder. The Save as dialog box is displayed when the user first saves the project. Figure 5-5 shows the project menu showing 'Save Project' command option.

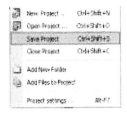

Figure 5-5: Project menu showing 'Save Project' command option

For a project, the full list of project folders and files, as well as the compiler/linker and debugger settings are saved. We can also save a project by right-clicking its name in the workspace window and selecting save project in the context menu.

Shortcuts:

Toolbar: No

Keys: SHIFT+CTRL+S


### 5.1.3.4 Closing a project

Use this command to close all windows containing the active project files. Aspire will suggest that we save changes to our project before it closes the associated files. If we close a project without saving, we will lose any changes made since the last time the project was saved. Before closing an untitled project, the Save Project dialog box is displayed by aspire and suggests that we name and save the project file. We can also close a project by right-clicking its name in the workspace window and selecting Close Project in the context menu.

Figure 5-6 shows project menu showing 'Close Project' command option.

Shortcuts:

Toolbar: No

Keys: SHIFT+CTRL+W or Delete when project is selected in workspace list.

Figure 5-6: Project menu showing 'Close Project' command option


### 5.1.3.5 Removing a project

We cannot remove an active/current project. To remove it, first make it inactive: right click the project in the workspace window and select Close Project in the context menu, or simply switch to another project. To remove a project from our workspace, once the project is inactive, right click the project in the workspace window select the Remove item in the context menu as shown in figure 5-7.

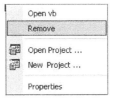

Figure 5-7: Project menu showing 'Remove' command option

We can add this project back to the workspace at any time by opening it.

### 5.1.3.6 Switching between projects

Our workspace may contain several projects (already added to workspace or created). To make a project active (current) double-click its name in the workspace window or select Open in its context menu shown by right-clicking its name. A project will be moved to the top of the workspace window, and the project folders tree will be shown. All project windows that were open the last time this project was in use will be restored.

### 5.1.3.7 Project settings

Every project has a unique set of settings. These settings are used to manage the project, they control how source files are compiled, built, and debug information generated. Along with common settings, we can define individual settings for each of the source files (if the compiler module allows this) using the file Properties. All of these settings are subject to the compiler module used with the project. After creating of a new project or opening the existing one, the settings for the project become available. To edit project settings, under the Project menu select Project settings as shown in figure 5-8, or use the Settings item in the context menu for your project in the workspace window.

Shortcuts:

Toolbar: No

Keys: ALT+F7

Figure 5-8: Project menu showing 'Project Settings' command option

### 5.1.3.8 Compiler settings

Every project has associated with it the compiler/linker modules that are used to build the source files. We can associate a stand-alone file in a project with any compiler supported by the module by right-clicking a file in the workspace window and choosing Properties in the context menu, individual compiler settings for a file can also be defined here (if the compiler module allows this).

We can also modify the settings for the compiler/linker module for the current project. These settings are usually concerning the paths to executable modules, libraries essential to use a compiler. These settings are applied to all projects that use this compiler/linker module.

After creating of a new project or opening the existing one, the settings for compiler module become available. To edit settings, select the Compiler settings command from the Build menu.

The settings for a specific project, concerning the way the compiler module manages the source files, compiles them, builds the object code, generates and uses the debugging information are found in the Project settings item of the Project menu. These settings will apply to any source file of any project that is preceded with the compiler.

To assign specific compiler settings for a separate file (if the compiler module allows this), use the file Properties. The compiler settings are specific to each of the installed compiler.

Shortcuts:

Toolbar: No

Keys: None

### 5.1.4 Managing files and folders in a project

### 5.1.4.1 Adding and removing files

To add a file to a project folder, first execute New (to create a new text file) or Open (to use an existing file) command from the File menu. Save a new file with the Save As command. Select Insert command as shown in figure 5-9 by right-clicking on the file in the editor, and insert the file into the current project.

Figure 5-9: Project menu showing 'Insert' command option

This will add the file to the current project. Aspire will add the file to a virtual folder and assign it a function in project (dictated by the file extension). The file can be dragged and dropped to other virtual folders in the project, or change its function. Alternatively, we can add an existing file to a folder by right-clicking the selected folder in the workspace window, and selecting the Add files to folder command as shown in figure 5-10, or using Add Files to Project under the Project menu item.

Figure 5-10: Dialog box showing 'Add Files to Folder' command option

To remove a file from the project, or to edit the file Properties, use the context menu by right-clicking the file in the workspace window, and select the desired command.

### 5.1.5 Compiling and Building

### 5.1.5.1 Compile

Use this command to compile the current active source file. The file will be processed by the compiler that is set in the properties of the file. By default, a compiler is assigned to a file depending on the file extension.

Shortcuts:

Toolbar:

Keys: Ctrl+F7

### 5.1.5.2 Build

Use this command to build an active project. All the source files that constitute the project will be compiled, if the existing output object files are found to be out of date (or no object file(s) found). This decreases the compilation overhead by processing only the sources that need to be processed. The compiler used is that defined by the file Properties option. To recompile all the source files creating new object files, use the 'Rebuild All' command.

Keys: Ctrl+Shift+F7 (for 'Rebuild All' command).

The object modules from the different sources will then be linked together to make a single output binary file, ready to be loaded in the target device for debugging.

Shortcuts:

Toolbar:

Keys: F7

### 5.1.5.3 Rebuild All

Use this command to deliberately recompile and link (if required) all the source files in the project, and build the output binary file without regard of the current status of the previously compiled object files.

Shortcuts:

Toolbar: None

Keys: Ctrl+Shift+F7

### 5.1.5.4 Stop Build

Immediately interrupts the Build, Rebuild All, and Compile commands. In most cases, the output files for the project remain inaccessible, and debugging of such a project is not possible until the project is successfully built. Only those errors and messages that were received until the build was interrupted, will be present in the output window. There may be a delay in interrupting the build process, and in some cases the build process may not be interrupted.

Shortcuts:

Toolbar:

Keys: F7

### 5.1.5.5 The steps involved in Compiling and Building a project:

A.S.P.I.R.E. has enhanced tools to compile and build a project .The typical steps included are the following:

After creating or opening a project and inserting source files, we can open any of the source files in the project by double-clicking the file name in the workspace window. A file is a source file (i.e. a file that will be compiled while the project build operations) when it is marked as a source. This mark is set in the file context menu that is called by right-clicking a file in the workspace window. We must also specify a compiler/linker for this file (if it is other than the default one for project) with the file Properties in the same menu. Files that are not specified as source files are not processed by aspire. Each file of a project may be assigned a specific compiler. For instance, if the compiler module allows, the source .asm file may be processed by the assembler, and the source .c file may be processed by the C compiler. When we add a file to the project, the compiler module assigns the default compiler to each source file. If we are redefining the assignment with the file Properties menu, we must consider the resulting effect. After the editing is complete, we can compile a file/project with the compiler/linker selected. Under the Project menu, select the Project settings option to tune the way the compiler module manages, compiles, and builds the source files, and the generation/use of debug information.

To compile a single source file: Make its file window active (open it in editor or click inside its window) and select the Compile command from Build menu (or press Ctrl+F7), or right-click a source file name in the workspace window and select Compile in the context menu as shown in figure 5-11.

Figure 5-11: Dialog box showing 'Compile' command option

The file will be processed by the compiler defined for the file. The compilation statistics and error/messages summary will be shown in the Output window, the Build tab as shown in figure 5-12.

102

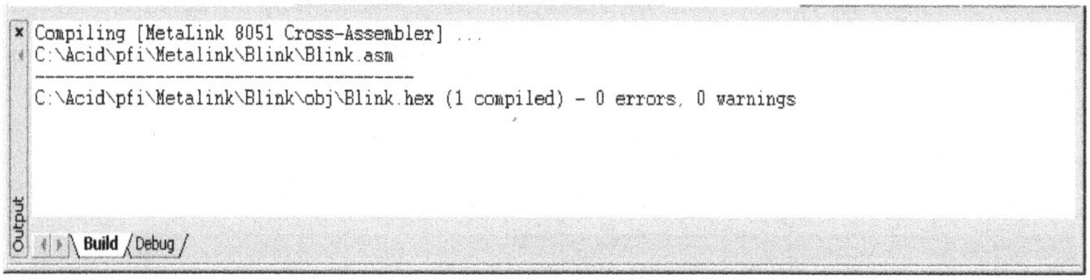

```
x  Compiling [MetaLink 8051 Cross-Assembler] ...
◄  C:\Acid\pfi\Metalink\Blink\Blink.asm
   -------------------------------------------
   C:\Acid\pfi\Metalink\Blink\obj\Blink.hex (1 compiled) - 0 errors, 0 warnings
```

◄ ► \ **Build** / Debug /

Figure 5-12: Output window showing compilation statistics and error/messages summary on the Build tab

In case of any errors or warnings, double-click a line containing the error/warning message in the Output window (or press F4). This will open a source file editor window with the cursor set at the corresponding source line. This line will be also marked with a blue dot to the left of it as shown in figure 5-13.

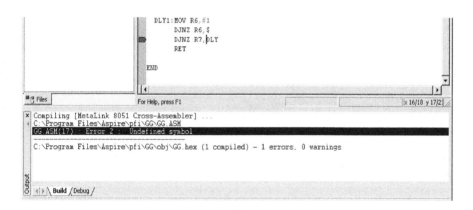

Figure 5-13: Output window showing error message/warning and the corresponding line marked with a blue dot to the left of it

To compile all the source files and link them into the single executable (binary) program body (if necessary): Select Build from Build menu (or press F7), or right-click a project file name in the workspace window and select Build context menu as shown in figure 5-14.

Figure 5-14: Dialog box showing 'Build' command option

Aspire will determine which of the files are to be recompiled and process only them. If an object file is up to date with the source file, no new compilation is applied. This decreases overhead by processing only the sources that need to be processed. If we wish to deliberately recompile all the files in the project, select Rebuild All command from Build menu as shown in figure 5-15 (or press Ctrl+Shift+F7). This will delete all the old object files, compile new ones, and then link them into the single executable (binary) program body.

Figure 5-15: Dialog box showing 'Rebuild All' command option

After our project is built without errors, we can then debug our program by starting debug. On finishing debugging, switch off debugging mode, and continue the edit/building/debugging development cycle until our program is completely functional and bug free. The output executable/binary file(s) produced by compiler/linker is in general the summary of our work.

**5.2 Software design**

**5.2.1 Flowchart for the calculation of blood oxygen saturation and the heart rate**

**5.2.1.1 Blood oxygen saturation:** Blood oxygen saturation is defined as the amount of oxygen dissolved in the blood. Saturation of peripheral oxygen ($SpO_2$) is an estimation of the oxygen saturation level usually measured with a pulse oximeter device. It can be calculated according to the following formula:

$$SpO_2 = HbO_2 \times 100\% \, / \, Hb + HbO_2$$

where $SpO_2$ is the percentage of blood oxygen saturation

$HbO_2$ is the amount of oxygenated hemoglobin

$Hb$ is the amount of deoxygenated hemoglobin

$Hb + HbO_2$ is the total hemoglobin

The relative absorption of light by oxyhemoglobin ($HbO_2$) and deoxyhemoglobin ($Hb$) is processed by the device and an oxygen saturation level is reported. Blood oxygen saturation is interpreted with the following table:

Table 5-1: $SpO_2$ Interpretation

| $SpO_2$ Reading (%) | Interpretation |
|---|---|
| 95-100 | Normal |
| 91-94 | Mild Hypoxemia |
| 86-90 | Moderate Hypoxemia |
| <85 | Severe Hypoxemia |

where hypoxemia is defined as decreased partial pressure of oxygen in blood and low oxygen availability to the body or an individual tissue or organ.

**5.2.1.2 Heart rate:** With each heart beat, the heart contracts and there is a surge of arterial blood, which momentarily increases arterial blood volume across the measuring site. This results in more light absorption during the surge. If light signals received at the photodiode are looked at 'as a waveform', there should be peaks with each heartbeat and troughs between heartbeats.

A newborn's heart rate is typically around 120 beats per minute (bpm). A heart rate in the vicinity of 70 beats per minute (bpm) is considered normal for an adult. When a person enters his golden years, the heart rate slows to approximately 50 bpm. When exercising, the heart rate may double. Accounting for all of this data, to say, 50 to 200 bpm are considered good readings for the heart rate.

The flowchart for the calculation of blood oxygen saturation and pulse rate is shown in figure 5-16.

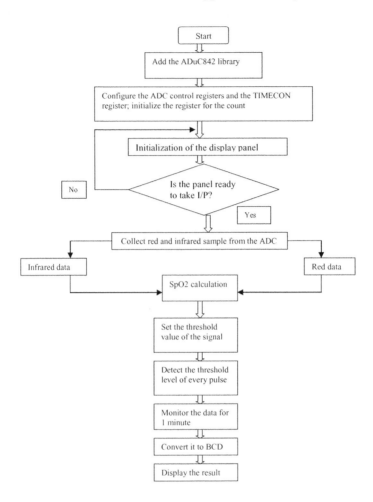

Figure 5-16: Flowchart for the calculation of blood oxygen saturation and heart rate

The ASCII code table used in programming is shown in table 5-2.

Table 5-2: ASCII code table

| Dec | Hex | Char | Dec | Hex | Char | Dec | Hex | Char | Dec | Hex | Char |
|---|---|---|---|---|---|---|---|---|---|---|---|
| 0 | 00 | Null | 32 | 20 | Space | 64 | 40 | @ | 96 | 60 | ` |
| 1 | 01 | Start of heading | 33 | 21 | ! | 65 | 41 | A | 97 | 61 | a |
| 2 | 02 | Start of text | 34 | 22 | " | 66 | 42 | B | 98 | 62 | b |
| 3 | 03 | End of text | 35 | 23 | # | 67 | 43 | C | 99 | 63 | c |
| 4 | 04 | End of transmit | 36 | 24 | $ | 68 | 44 | D | 100 | 64 | d |
| 5 | 05 | Enquiry | 37 | 25 | % | 69 | 45 | E | 101 | 65 | e |
| 6 | 06 | Acknowledge | 38 | 26 | & | 70 | 46 | F | 102 | 66 | f |
| 7 | 07 | Audible bell | 39 | 27 | ' | 71 | 47 | G | 103 | 67 | g |
| 8 | 08 | Backspace | 40 | 28 | ( | 72 | 48 | H | 104 | 68 | h |
| 9 | 09 | Horizontal tab | 41 | 29 | ) | 73 | 49 | I | 105 | 69 | i |
| 10 | 0A | Line feed | 42 | 2A | * | 74 | 4A | J | 106 | 6A | j |
| 11 | 0B | Vertical tab | 43 | 2B | + | 75 | 4B | K | 107 | 6B | k |
| 12 | 0C | Form feed | 44 | 2C | , | 76 | 4C | L | 108 | 6C | l |
| 13 | 0D | Carriage return | 45 | 2D | − | 77 | 4D | M | 109 | 6D | m |
| 14 | 0E | Shift out | 46 | 2E | . | 78 | 4E | N | 110 | 6E | n |
| 15 | 0F | Shift in | 47 | 2F | / | 79 | 4F | O | 111 | 6F | o |
| 16 | 10 | Data link escape | 48 | 30 | 0 | 80 | 50 | P | 112 | 70 | p |
| 17 | 11 | Device control 1 | 49 | 31 | 1 | 81 | 51 | Q | 113 | 71 | q |
| 18 | 12 | Device control 2 | 50 | 32 | 2 | 82 | 52 | R | 114 | 72 | r |
| 19 | 13 | Device control 3 | 51 | 33 | 3 | 83 | 53 | S | 115 | 73 | s |
| 20 | 14 | Device control 4 | 52 | 34 | 4 | 84 | 54 | T | 116 | 74 | t |
| 21 | 15 | Neg. acknowledge | 53 | 35 | 5 | 85 | 55 | U | 117 | 75 | u |
| 22 | 16 | Synchronous idle | 54 | 36 | 6 | 86 | 56 | V | 118 | 76 | v |
| 23 | 17 | End trans. block | 55 | 37 | 7 | 87 | 57 | W | 119 | 77 | w |
| 24 | 18 | Cancel | 56 | 38 | 8 | 88 | 58 | X | 120 | 78 | x |
| 25 | 19 | End of medium | 57 | 39 | 9 | 89 | 59 | Y | 121 | 79 | y |
| 26 | 1A | Substitution | 58 | 3A | : | 90 | 5A | Z | 122 | 7A | z |
| 27 | 1B | Escape | 59 | 3B | ; | 91 | 5B | [ | 123 | 7B | { |
| 28 | 1C | File separator | 60 | 3C | < | 92 | 5C | \ | 124 | 7C | | |
| 29 | 1D | Group separator | 61 | 3D | = | 93 | 5D | ] | 125 | 7D | } |
| 30 | 1E | Record separator | 62 | 3E | > | 94 | 5E | ^ | 126 | 7E | ~ |
| 31 | 1F | Unit separator | 63 | 3F | ? | 95 | 5F | _ | 127 | 7F | ▯ |

## 5.3 Screen shots

Figure 5-17: Workspace for writing the program

Figure 5-18: Program showing zero errors and zero warnings

108

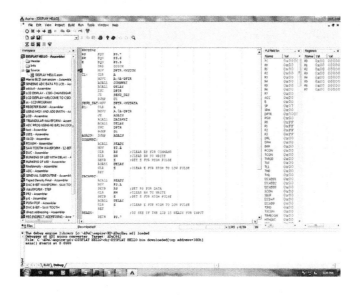

Figure 5-19: Compilation of the program

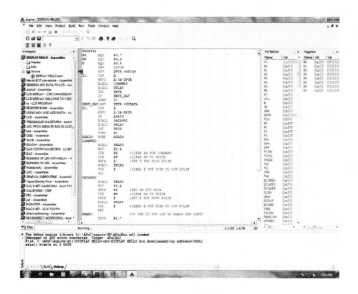

Figure 5-20: Running of the program

109

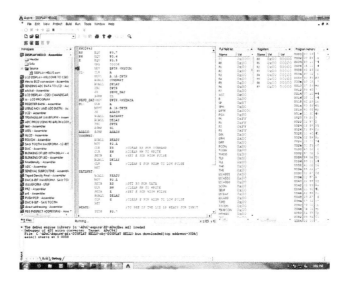

Figure 5-21: Results of the program

110

# CHAPTER 6

## RESULTS

### 6.1 Readings

The readings which I got after running the program were:

Blood oxygen saturation: 97%

Pulse rate: 79 BPM (beats per minute)

### 6.2 Results

The blood oxygen saturation and the heart rate have been displayed on the OLED display as shown in figure 6-1. The waveform for the heart beat has also been displayed.

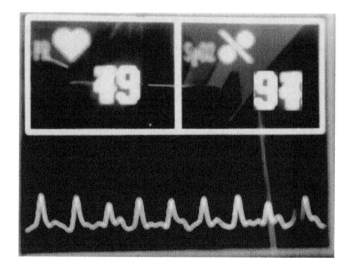

Figure 6-1: Display of heart rate and blood oxygen saturation along with the waveform

### 6.3 Conclusion

The pulse oximeter is a very useful device for monitoring patients during anesthesia, intensive care, emergency departments, general wards or those with conditions such as asthma. The work, in essence was carried out as an application of biomedical instrumentation. The readily available pulse oximeter sensor was used to extract the signal from the finger. Different signal conditioning equipments like the filter and amplifier helped to strengthen the signal which was low in amplitude and frequency. The system uses the ADuC842 as a microcontroller and has very low power consumption. The system has attractive features to measure the $SpO_2$ and the pulse rate. The programming is done in assembly language. The display panel used is OLED which is a new technology and has high pixel density. Use of microcontroller, analog to digital converters, different hardware and proper interfacing among these devices led to the development of the 'Pulse Oximeter'. The system thus designed was a standalone system that can be used just by providing it with the power supply. The system works satisfactorily displaying the $SpO_2$ level and heart rate in the desired range. However there does remain certain room for future enhancements. Hence it can be concluded that the objective of this work has been achieved and I have been successful in developing the "Pulse Oximeter".

### 6.4 Future Scope

The future scope of the pulse oximeter made is that it can be made wireless while implementing wireless transmitter and receiver. Alarm can also be provided to reduce morbidity and increase monitoring. Alarms occur if the machine is faulty, the patient-machine connection is faulty or if the patient develops a fault.

# REFERENCES

The references include the following:

1.  Cho Zin Myint, Nader Barsoum, Wong Kiing Ing, "Design a medicine device for blood oxygen concentration and heart beat rate", Global Journal on Technology and Optimization, June 2010.

2.  Maziar Tavakoli, Student Member, IEEE; Lorenzo Turicchia, and Rahul Sarpeshkar, Senior Member, IEEE, "An Ultra-Low-Power Pulse Oximeter Implemented with an Energy-Efficient Transimpedance Amplifier", IEEE transactions on Biomedical circuits and systems, VOL. 4, NO. 1, February 2010.

3.  Pawan K. Baheti, Harinath Garudadri, "An Ultra low power pulse oximeter sensor based on compressed sensing", IEEE Computer Society, 2009.

4.  Susannah Fleming, Lionel Tarassenko, Matthew Thompson, and David Mant, "Non-invasive Measurement of Respiratory Rate in Children Using the Photoplethysmogram", 30th Annual International IEEE EMBS Conference, August 2008.

5.  Stephan Reichelt_, Jens Fiala, Armin Werber, Katharina Förster, Claudia Heilmann, Rolf Klemm, and Hans Zappe, "Development of an Implantable Pulse Oximeter", IEEE transactions on Biomedical Engineering, VOL. 55, NO. 2, February 2008.

6.  Guowei Di, Xiaoying Tang, Weifeng Liu, " A Reflectance Pulse Oximeter Design Using the MSP430OF149", IEEE/ICME International Conference on Complex Medical Engineering, 2007.

7.  Hassan Deni, Diane M. Muratore, Robert A. Malkin, "Development of a Pulse Oximeter Analyzer for the Developing World", IEEE, 2005.

8.  P. C. Branche, W. S. Johnston, C. J. Pujary, and Y. Mendelson, "Measurement Reproducibility and Sensor Placement Considerations in Designing a Wearable Pulse Oximeter for Military Applications," 30th Annual Northeast Bioengineering Conference, 2004.

9.  U. Anliker, et al., "AMON: A Wearable Multiparameter Medical Monitoring and Alert System", IEEE Transactions on Information Technology in Biomedicine, 2004

10. Panayiotis A. Kyriacou, Sarah Powell, Richard M. Langford, and Deric P. Jones, "Esophageal Pulse Oximetry Utilizing Reflectance Photoplethysmography", IEEE transactions on Biomedical Engineering, VOL. 49, NO. 11, November 2002.

11. Juan M. Lopera, Juan Diaz, Miguel J. Pneto and Fernando Nuno, "Pulse oximeter for homecare", IEEE Proceedings of the Second Joint EMBSBMES Conference, October 2002.

12. S. Rhee, B.H. Yang and H. Asada, "The Ring Sensor: A New Ambulatory Wearable Sensor for Twenty Four Hour Patient Monitoring." Proc. of the 20th Annual International Conference of the IEEE Engineering in Medicine and Biology Society, Hong Kong, Oct. 1998.

13. P,B. Crilly, et al., "An Integrated Pulse Oximeter System for Telemedicine Applications," Proc. of the IEEE Instrumentation and Measurements Technical Conference, Ottawa, Canada, May 19-21, 1997

# BIBLIOGRAPHY

1. Data Conversion Handbook by Analog Devices.

2. Op-Amps and Linear Integrated Circuits, Third Edition, by Ramakant A. Gayakwad.

3. Op Amp Applications Handbook, by Walt Jung.

4. Organic Light Emitting Devices: Synthesis, Properties and Applications, by Klaus Müllen, Ullrich Scherf.

5. Switching Transistor Handbook, by William D. Roehr.

6. The 8051 Microcontroller and Embedded systems, by Muhammad Ali Mazidi and Janice Gillispie Mazidi.

Made in the USA
Monee, IL
28 January 2021